Revised Printing
Practical Lab Manual for

PHYSICAL
geology

Kendall Hunt
publishing company

Nabil Kanja

Moraine Valley Community College

Cover image © 2015 Shutterstock, Inc.
Interior images courtesy of Nabil Kanja unless otherwise stated.
All National Geographic maps used throughout were created using TOPO!
Software © 2011 National Geographic Maps.

Kendall Hunt
p u b l i s h i n g   c o m p a n y

www.kendallhunt.com
Send all inquiries to:
4050 Westmark Drive
Dubuque, IA 52004-1840

Contents

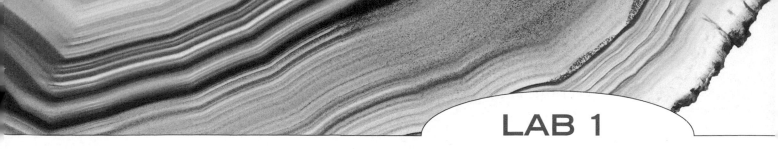

MINERALS AND CRYSTAL GROWTH

Minerals could form in more than one way, from different processes. Some of these processes include the following:

- Evaporation and precipitation from solution like evaporites (halite and gypsum)
- Freezing, like formation of ice from water
- Cooling of magma, by crystallization in igneous rocks
- Change of pressure and temperature (increase of pressure and temperature in metamorphic rocks and metamorphism processes)

Figure 1.1

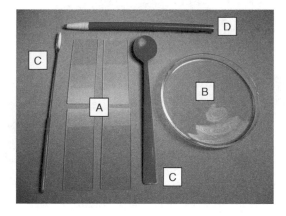

Figure 1.2

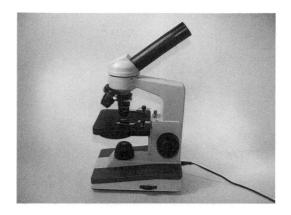

Figure 1.3

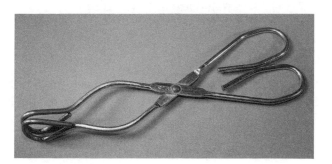

Figure 1.4

Before starting with experiments you will need the following tools (or some of them):

■ Hot plate. See Figure (1.1)

■ Eight glass slides. See Figure (1.2-A)

■ Two Petri dishes. See Figure (1.2-B)

■ Spoon(s) or spatula. See Figure (1.2-C)

■ Grabber(s). See Figure (1.4)

■ Wax pen. See Figure (1.2-D)

■ Masking tape.

■ Simple microscope. Figure (1.3)

■ Solutions.

■ Watch.

PART (A)

1.A) There are four solutions and these solutions are as follows:

● Sodium Chloride

● Sodium Nitrate

● Potassium Aluminum Sulfate

● Copper Accetate

2.A) Bring four slides and put two drop from each solution (using dropper) on each slide. Make sure that you do not touch the solutions with hand. Label the slides using masking tape and marker or wax pencil. For example, Sodium Chloride will be (A), Sodium Nitrate will be (B), Potassium Aluminum Sulfate will be (C), and Copper Accetate will be (D). Allow the solutions on each slide to evaporate or dry naturally, so do not use any heaters (hot plates), and let the air in room do the evaporation and this will take time, approximately (45) minutes. Remember evaporation is not melting. Evaporation is the process where liquid changes to gas, whereas melting is the process where solid changes to liquid, and in both processes heat is added, naturally or not naturally.

3.A) After complete evaporation, put each of the four slides on the stage of the microscope supplied and try to draw the crystals that formed (if there are crystals) in the following squares. Pay attention to shape of crystal(s), direction, connected or separated, size and whether crystal(s) are individual or aggregates.

Figure 1.5 Sodium Chloride (A)

Figure 1.6 Sodium Nitrate (B)

Figure 1.7 Potassium Aluminum Sulfate (C)   **Figure 1.8** Copper Accetate (D)

4.A)   Which solution(s) is/are showing cubic crystals?

5.A)   Which solution(s) is/are showing aggregates (group) of crystals?

6.A)   What process or processes produced these crystals (if any)? Explain.

7.A)   Does each solution give a unique crystal form (shape)?

8.A)   Can minerals be identified from crystal form (shape) alone? Be careful when you answer this question.

9.A)   Try to match the following description of crystals to Figures (1.5)–(1.8):

- Green (sometimes blue) crystals that are rhombic in shape. These crystals take time to form (slow growing) and may produce incomplete or not well-formed crystals: _____
- Colorless (clear) crystals which are cubic in shape. These crystals take time to form however nice shapes may be obtained if given time and right conditions: _____
- Colorless (clear) crystals which are octahedral (like diamond) or cubic in shape. These crystals form in relatively short period of time: _____
- Colorless (clear) crystals that are hexagonal in shape and the crystals look like calcite and are challenging to grow: _____

PART (B)

1.B)   Bring hot plate and plug it then put the setting on (300). See Figure (1.1). Be careful not to touch the hot plate and make sure that the hot plate is not interfering with you or others.

2.B)   Bring four new slides and put two drop from same previous solution (using dropper) on each slide. Make sure that you do not touch the solutions with hand. Again, label the slides using masking tape and marker or wax pencil that is, Sodium Chloride will be (A1), Sodium Nitrate will be (B1), Potassium Aluminum Sulfate will be (C1), and Copper Accetate will be (D1).

- Take slide (A1), Sodium Chloride, and carefully put it on the hot plate. Do not put the slide completely on the hot plate but half of the slide and only leave it for (30) seconds then take the slide from the hot plate and put it on the side to cool.
- Take slide (B1), Sodium Nitrate, and carefully put it on the hot plate. Do not put the slide completely on the hot plate but half of the slide and only leave it for (1.5) minutes to (2) minutes and then take the slide from the hot plate and put it on the side to cool.
- Take slide (C1), Potassium Aluminum Sulfate, and carefully put it on the hot plate. Do not put the slide completely on the hot plate but half of the slide and only leave it for (20) seconds and then take the slide from the hot plate and put it on the side to cool.
- Take slide (D1), Copper Accetate, and carefully put it on the hot plate. Do not put the slide completely on the hot plate but half of the slide and only leave it for one minute and (20) seconds and then take the slide from the hot plate and put it on the side to cool.

3.B) After all the slides cooled, put each slide at a time on the stage of the microscope and drew what you see in the following boxes:

Figure 1.9 Sodium Chloride (A1)   **Figure 1.10** Sodium Nitrate (B1)

Figure 1.11 Potassium Aluminum Sulfate (C1)   **Figure 1.12** Copper Accetate (D1)

PART (C)

1.C) Put some grains of crystalline Phenyl Salicylate (alternative to magma) on Petri dish and then put the dish on hot plate using the supplied grabbers. Make sure that the setting for the hot plate is always the same during all steps of experiments (300) and is not high. Be careful, do not touch Phenyl Salicylate and be careful when using the hot plate.

2.C) Allow the Phenyl Salicylate to melt (not evaporate and remember melting is changing from solid state to liquid state). The melting will take approximately (15) seconds using (300) setting.

3.C) Take the dish that contains the melting Phenyl Salicylate using supplied grabbers (never use your hands, the dish will be very hot) and put it on the side so that the Phenyl Salicylate cools with time.

4.C) After cooling, put the dish on the stage of the microscope using grabbers (never use your hands) and draw the crystal(s) that may form (if any). Draw what you see in the box of Figure (1.13).

Figure 1.13

5.C) Did you get individual crystal(s) or aggregates of crystals?

6.C) Is/are the crystal(s) small or large one(s)?

7.C) Was rate of cooling fast or slow? Slow, means more than (10) minutes; fast means less than (10) minutes (this is relative).

PART (D)

1.D) Put some grains of crystalline Phenyl Salicylate on different dish and then put the dish on hot plate. Be careful, do not touch Phenyl Salicylate and be careful when using the hot plate.

2.D) Allow the Phenyl Salicylate to melt and as soon as the Phenyl Salicylate starts melting put 4 to 5 "seeds" to the melt in the dish using the spoon supplied. Allow the "seeds" to melt. The melting may take (15) seconds or more using (300) setting.

3.D) Take the dish that contains the melting Phenyl Salicylate and the "seeds" using the supplied grabbers (again, never use your hands, the dish will be very hot) and put it on the side so that the mixture of Phenyl Salicylate and "seeds" cools with time.

4.D) After cooling, put the dish on the stage of the microscope and draw the crystal(s) that may form (if any). Draw what you see in box of Figure (1.14).

Figure 1.14

5.D) Did you get individual crystal(s) or aggregates of crystals?

6.D) Is/are the crystal(s) small or large one(s)?

7.D) Was rate of cooling fast or slow?

8.D) Compare Figures (1.13) and (1.14). Which figure out of the two figures show larger crystal(s)?

9.D) Mention the reason(s) for your answer for question (8.D); that is say why.

10.D) Did the addition of "seeds" to Phenyl Salicylate in part (D) increase the size of crystal(s) compared to size of crystal(s) in part (C)?

11.D) From part (A), part (B), part (C), and part (D) explain how crystals (minerals) formed in these previous experiments?

12.D) From part (A), part (B), part (C), and part (D) mention some factors that control and influence the size and shape of crystals.

PHYSICAL PROPERTIES OF MINERALS

Before introducing physical properties (characteristics), it is important to know that color, streak, luster, and diaphaneity can be put under optical properties since light is involved. Also cleavage, fracture, hardness, and tenacity can be put under mineral strength since it depends on the types of bonds and the strength of such bonds.

Now, in order to identify minerals, that is, to name them, we must have some knowledge about practical physical properties of such minerals and these include the following:

- Color
- Cleavage
- Hardness
- Luster
- Streak
- Fracture
- Crystal form
- Crystal habit
- Other properties

The tools that are used for mineral identification are simple and available in most hardware stores. Figure (2.1) shows these tools and they include streak plate (white), glass plate, penny (one cent), magnifying glass, magnet and bottle of diluted hydrochloric acid (HCl).

Figure 2.1

COLOR

Color cannot be used to identify minerals since some minerals have more than one color like quartz. Quartz could be white, colorless, smoky, rosy, black, and other colors. Look at Figure (2.18-1 to 6).

The reason some minerals have more than one color is due to the following:

- Inclusions: are minerals within other minerals like rutile in quartz.
- Impurities: are ions found in minerals like iron (Fe) in amethyst giving it the purple color.
- Defects in the structure of mineral.
- Radiation like smoky quartz.
- Oxidation state. For example iron ion (Fe^{+2}) gives green color whereas (Fe^{+3}) gives red to brown color in some minerals containing iron.

CLEAVAGE

Definition

If the weak bonds between atoms are broken, then a plane or surface of weakness will or could develop, along them, we call these planes or surfaces, *cleavage*. So cleavage planes are planes of weakness.

For example, if you put stress (knife blade) between atoms of basal mineral like mica in specific direction as shown in Figure (2.2) then bonds between atoms are broken along planes, these planes are known as cleavage planes.

© Tyler Boyes/Shutterstock, Inc. Carrie

Figure 2.2

How do the cleavage planes look like?

Usually, they are like flat smooth surfaces that are shiny. Sometimes cleavage planes look like steps (as if they are going in and out). Look at Figure (2.3).

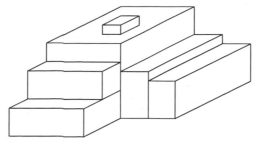

Figure 2.3

How can you tell if a mineral has a cleavage?

Usually, if you took a sample and brook it, the small fragments produced, will have same cleavage planes that are parallel to the original ones, for example, if you took the mineral galena which has a cubic shape, look at Figure (2.4), and brook it to pieces, the small pieces will also show cleavage planes similar to the original ones and parallel to them.

Figure 2.4

As a conclusion, we can say from the previous example that if mineral was broken to small fragments and the fragments were having same shape like the original one then that specific mineral will have cleavage; however, if this did not happen (shapes are different) then these surfaces are not cleavage (planes of weakness) but are fractures.

Of course, you are not allowed to break minerals so either the instructor tells you the names of minerals that have cleavage (or do not have cleavage) or simply you can tell by practicing and training.

Furthermore, remember that although some minerals show shiny flat planes, these planes are not cleavage planes but what are known as crystal faces, and it is important to say that usually cleavage planes run parallel to crystal faces, but not always.

Some minerals could show flat and shiny surfaces but have *no cleavage* like quartz (Figure 2.18-1), pyrite (Figure 2.18-35), garnet (Figure 2.18-23), and corundum (Figure 2.18-39). Furthermore, the following minerals have no cleavage like chalcopyrite, granular olivine, hematite, magnetite, and most of native elements. The previous examples are only some of the minerals that have no cleavage. Also note that some forms of the same mineral may not show cleavage. For example, selenite (Figure 2.18-54), one form of gypsum, has cleavage whereas satin spar (Figure 2.18-53), another form of gypsum, shows no cleavage.

Description of Cleavage

Cleavage should be described and that includes the number of sets and angles between cleavage planes. The following are different types of cleavage:

■ *Basal cleavage* is one set of cleavage that is parallel to the base of the mineral like mica, talc, chlorite, and graphite. Sometimes known as pinacoidal cleavage. Look at Figure (2.5).

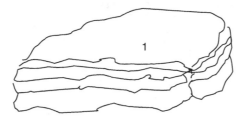

Figure 2.5 One set of cleavage or one plane of cleavage since all surfaces are parallel to each other, like mica and graphite, but many surfaces of cleavage

■ **Prismatic cleavage** refers to two sets of cleavage but elongated like pyroxene and amphibole. Look at Figure (2.6).

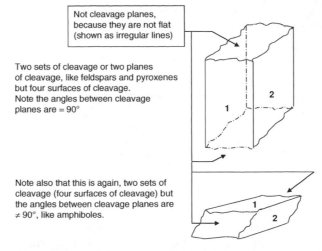

Not cleavage planes, because they are not flat (shown as irregular lines)

Two sets of cleavage or two planes of cleavage, like feldspars and pyroxenes but four surfaces of cleavage. Note the angles between cleavage planes are = 90°

Note also that this is again, two sets of cleavage (four surfaces of cleavage) but the angles between cleavage planes are ≠ 90°, like amphiboles.

Figure 2.6

■ *Isometric cleavage* is same as *cubic cleavage* and refers to three sets of cleavage planes with 90° angles like halite and galena. Look at Figure (2.7).

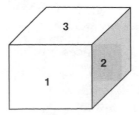

Figure 2.7 Three sets of cleavage or three planes of cleavage, like halite and galena, but six surfaces of cleavage. Note that the angle between cleavage planes is = 90°

■ *Rhombohedral cleavage* (mineral have the shape of lopsided cube or collapsed rectangle) refers to three sets of cleavage planes that are not 90° like calcite and gypsum. Look at Figure (2.8).

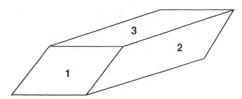

Figure 2.8 Three sets of cleavage or three planes of cleavage and six surfaces of cleavage, like calcite and gypsum. Note that the angles between cleavage planes are ≠ 90°

■ *Octahedral cleavage* refers to four sets of cleavage planes (eight surfaces) and the angles between the cleavage planes are not 90° like fluorite. Look at Figure (2.9).

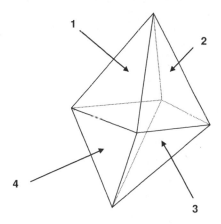

Figure 2.9 Four sets of cleavage, or four planes of cleavage and eight surfaces of cleavage, like fluorite

■ *Dodecahedral cleavage* refers to six sets of cleavage planes (12 faces) and the angles between cleavage planes are not 90° like sphalerite. Look at Figure (2.10).

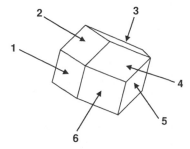

Figure 2.10 Six sets of cleavage, or six planes of cleavage and 12 surfaces of cleavage, like sphalerite

Important Notes to Consider

■ It is very important to say, as noticed earlier, from the above figures that the sets of cleavage are same as planes of cleavage, which are referring to group of cleavage *surfaces* that are *parallel* to each other and that are in a *specific direction*. Also note that those surfaces that are parallel to each other in a specific direction are considered one plane or set of cleavage; therefore, a cube like the mineral halite has three sets or planes of cleavage but six surfaces of cleavage.

Note also, that the above angles are the angles between cleavage planes that *intersect* (adjacent to each other) and such angles could be 90° or different than 90°.

■ Cleavage direction refers to orientation of different sets of parallel cleavage planes. It is same as saying: "one set of cleavage or two sets of cleavage with 90° etc." So as described before, you have to mention how many sets and the angles between them if it is more than one set of cleavage.

■ Although some minerals do have cleavage however sometimes it is not possible to see it (mineral could be broken) and what you can see is small or very small parts of cleavage planes that are flat and shiny so try to observe and identify these small surfaces. If you do not see flat and shiny surfaces then look for steps as mentioned earlier.

■ Cleavage can be described also by using the terms excellent, good, and poor cleavage as follows:

• Excellent and good cleavage refers to large flat and even surfaces, and this means that you can see or detect the cleavage plane(s). Usually, these planes are shiny because they reflect large amount of light.

• Poor cleavage refers to small, flat uneven surfaces, and this means that the mineral have cleavage but is difficult to see and detect or it is not showing cleavage planes (you can't see them) although the mineral do have cleavage. The cleavage plane(s) are less shiny because only small amount of light is reflected whereas the rest or remaining rays of the light are reflected in different directions (scattered) from fracture surfaces.

• Remember fractures are not flat and shiny surfaces because they reflect light in different directions (scattered light).

■ Cleavage could develop along the weakest bond, however if the bonds have the same strength and the bonds are not arranged in a specific pattern then fractures will develop but if the bonds have all same strength and arranged in a specific pattern that will allow the cleavage plane(s) to develop then cleavage plane(s) will develop.

HARDNESS

Is the ability of the mineral to resist abrasion (scratching).

The hardness test, or as known as Mohs scale, look at Figure (2.11), can be performed like this:

1. Select a fresh surface of mineral, a surface that is not weathered or oxidized.
2. Start the test by scratching the mineral with the following tools in this order; first use fingernail (hardness = 2.5) then coin (hardness = 3.5) then glass (hardness = 5.5) then streak plate (hardness = 6.5).

In addition to the above tools, minerals are given to give more precise hardness of the mineral sample:

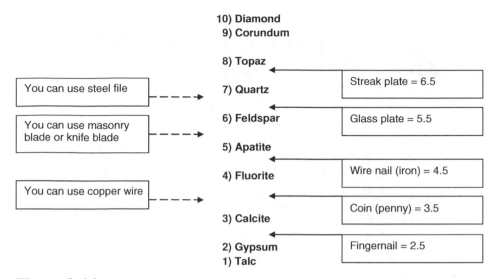

Figure 2.11

3. For example, if you have a sample of mineral of unknown hardness, use first your fingernail, if the fingernail scratches the sample then the hardness of the sample is less than 2.5. If fingernail did not scratch the sample then use the next tool (coin) the same way. If coin scratches the sample then you stop and hardness of sample will be between 2.5 and 3.5, but if coin did not scratch the sample then you go to next tool (wire nail) and you repeat the above again.

4. Always know what is scratching what; is it the sample or the tools and minerals you use. Make sure that it is a real scratch by rubbing the scratch and see if it vanishes, if so then this is not scratch but powder of the sample or the tool.

5. Although, individual minerals, each have one value for hardness however some minerals show more than one. Best example is hematite; the hardness could range from 1 to 6. Other minerals, for example calcite shows different hardness values depending on directions used for scratching. Calcite have hardness of 3, however it could be 2 on one of the surfaces (directions).

6. The numbers in hardness test are relative. For example, although the hardness of corundum is one time harder than topaz and two times harder than quartz, however in reality it is two times as hard as topaz, and four times harder than quartz.

7. Hardness of tools like glass although some times is written 5.5; however, it could be between 5 and 5.5. Same thing with streak plate, its hardness could range between 6 and 7. Again the tools that you are using, including your fingernail, could vary, but not that much.

8. Hardness refers to the ability to be scratched and not the ability to break. For example, diamond is very hard however is easy to break where it shatters to pieces since it is brittle. Also remember that hardness is related to the strength of the bonds.

9. Mohs scale gives relative hardness; however, it is possible to determine absolute hardness by knowing relative hardness and by using the curve that is shown in Figure (2.12). For example, calcite has relative hardness of (3) but it is about (9) times harder than talc and that is considered absolute hardness.

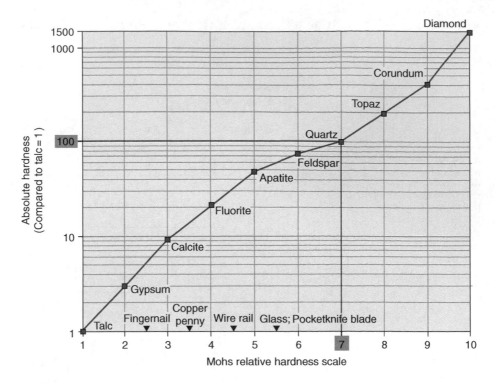

Figure 2.12

LUSTER

||||||||||||||||||||||||||||||||||||||||||||||||||||||||||||||||||||||||||||||||||||||||||||||||||||||||||||||||||||||||||||||||||||||||||||||||||||||||||||||||||||||||||||||||||||||||||

Luster refers to how the surfaces look like when it reflects the light, or simply, the appearance of a mineral surface in a reflected light.

There are two types of luster:

1. Metallic: the minerals have the look of metal surfaces like silver, iron, gold, and copper. Usually, minerals with metallic luster are heavy, opaque and give dark streak. Some examples of minerals showing metallic luster are pyrite, galena, graphite, and some hematite. Look at Figure (2.18-35, 38, 56).

   The word "shiny" could be applied to metallic luster but not all shiny minerals have metallic luster.

2. Nonmetallic: the minerals do not have metal appearance and look like glass or colored glass. Dull surfaces are also considered nonmetallic. Nonmetallic minerals are usually light colored, transparent to translucent and light in weight. The streak is usually colorless to very light color. Nonmetallic luster is further divided and given specific terms to describe it:

   a. Vitreous: the surface looks like a glass (e.g., quartz, calcite, feldspar, olivine, mica).
   b. Resinous: the surface looks like resin (e.g. garnet, sphalerite). Look at Figure (2.18-23, 40).
   c. Pearly: the surface looks like that of pearl (e.g., talc, aragonite, and dolomite). Look at Figure (2.18-25).
   d. Silky: the surface is very finely fibrous (e.g., serpentine or asbestos and satin spar, a variety of gypsum). Look at Figure (2.18-15, 53).

    e.   Adamantine: although close to vitreous, but is more shiny or fiery like diamond and some times, like olivine. Look at Figure (2.18-48).

    f.   Dull or earthy: not shiny or bright (e.g., kaolinite and bauxite). Look at Figure (2.18-24, 60).

Another type of luster can be added, and it is considered between the two of the above:

3. Sub metallic: when you feel that the mineral is in between the metallic and nonmetallic luster or part of the mineral is metallic, then it is OK to describe it as submetallic (e.g., some hematite).

STREAK

Is the color of the powder, the mineral leaves on the streak plate. Look at Figure (2.13). To determine the streak of a specific mineral, try to note the following:

1. Use the fresh surface of the mineral to get streak.
2. Always make sure of color of the powder for the specific mineral, especially when it is white or colorless (no color shown on streak plate), by rubbing the powder with your finger and noticing the color of the powder on your finger. Do this especially when you cannot see the color on the white porcelain plate.
3. Monocolor minerals like sulfur (different samples always have same color) have streak same as the color of mineral, whereas multicolor minerals (different samples of same mineral have differ-ent colors like hematite and quartz) will have streak different than the color of mineral.
4. Remember that the hardness of the unglazed porcelain plate (streak plate) is between 6 and 7 and any mineral harder than the streak plate will scratch the streak plate and the powder will be that of the streak plate and not of the mineral and in this case the streak of the mineral is unknown. The unknown streak of the mineral is determined from mineral identification table.
5. Be careful, and notice that the streak is not always as same as the color of the mineral.

Figure 2.13

FRACTURES

If mineral was broken to small fragments and the shape of the small fragments did not look like the shape of the original parent mineral then this specific mineral will have no cleavage and what will develop are random surfaces and these random surfaces are not shiny and not flat. These random surfaces are known as fractures.

Fractures will develop when the bonds between atoms are strong in all directions and hence fractures are breakage that forms a surface with no relationship to the internal structure of the mineral (not related to the bonds).

Different kinds of fracture patterns are observed:

- *Conchoidal fracture*—breaks along smooth curved surfaces resembling the interior surface of a shell like obsidian (rock), opal, flint, and sometimes quartz. Look at Figure (2.14-1, 2)
- *Fibrous and splintery*—similar to the way wood breaks, like satin-spar or chrysotile (type of serpentine). Look at Figure (2.14-3).
- *Hackly*—jagged fractures (saw toothed) with sharp edges. Some examples are copper and cast iron. Look at Figure (2.14-4)
- *Uneven or Irregular*—rough irregular surfaces, most minerals have this type of fracture. Look at Figure (2.14-5).
- *Even*—the fracture surface is nearly flat, like chert and kaolinite. Look at Figure (2.14-6).

Some same minerals with different samples could display different fractures and the terms used to describe the fracture will also depend on the student. For example, serpentine may show uneven/irregular fracture when the fibers are not seen; however, it may show fibrous/splintery fracture when these fibers are noticed.

It is important to say that a mineral can have both cleavage and fracture at the same time. For example if a mineral with cleavage was broken in a certain way then fractures are produced. All minerals (including those that have cleavage) have fractures.

Figure 2.14-1 Conchoidal fractures in obsidian (circles)

Figure 2.14-2 Conchoidal fractures in obsidian (half circles)

www.sandatlas.org/Shutterstock, Inc.

Figure 2.14-3 Fibrous fractures in chrysotile

Figure 2.14-4 Hackly fractures in copper

Figure 2.14-5 Uneven fractures in magnetite

Figure 2.14-6 Even fractures in kaolinite

CRYSTAL FORM

Is referring to the geometric shape of the mineral crystal or a mineral that develops as a crystal with a regular pattern of faces and angles between the adjoining faces, which are characteristic of a particular mineral. Minerals will display good crystal form only if they are allowed to grow in an unrestricted environment, which is rather rare in nature. Most of the time minerals are competing for space and grow into one another creating a mass of interlocking crystals. So, many times you will not observe a good crystal form and cannot use this property for identification.

Generally speaking, the crystal form means the set of crystal faces that have same shape and in crystallography, form is face or set of faces that are similar because they have the same arrangement of atoms and form is considered a sub classification of the crystal system.

CRYSTAL HABIT

Is the external shape of a crystal (Individual Crystals, I.C) or aggregate of crystals (Crystal Aggregates, C.A) and the relation of individual crystal faces to other crystal faces. Note that crystal aggregates could be small or fine crystals that look like grains. The following are some terms that are used to describe crystal habit:

■ *Acicular (IC):* Fine needle-like crystals. Those are closely packed, as in natrolite or manganite radiating or divergent from the center. Another example is pyrite.

■ *Bladed (IC):* Occurs as flat blade or thin rectangular shape, like kyanite.

■ *Fibrous (IC):* Consisting of fine thread-like strands (fibers), as shown by the variety of gypsum called satin-spar, and also by asbestos.

■ *Prismatic (IC):* Elongation of the crystal in one direction, as in the feldspars, the pyroxenes, and the common hornblendes.

■ *Granular (CA):* Look like grains or aggregates of grains that are either fine or coarse, like olivine.

■ *Dendritic (CA):* Plant like or leaf shape, like manganese oxide.

■ *Amygdaloidal (CA):* Almond-shaped crystals or grains of minerals that occupy the cavities or vesicles (gas holes) in lava flows.

SPECIAL PHYSICAL PROPERTIES

Some minerals show certain properties that are easily identified and hence it is easy to know the name of unknown mineral from that specific property:

Magnetism, when you bring the mineral magnetite to a magnet, the magnet will attract it, whereas the mineral hematite will not be attracted to a magnet, although sometimes it does. Look at Figure (2.15). The above information is not accurate enough and is used to differentiate between different minerals as a simple tool; however, there are minerals that could be attracted to magnet or even acting like a magnet. The following are some examples:

■ Magnetite: attracted to magnet and sometimes acts like magnet.

■ Lodestone: same as magnetite that is, it is attracted to magnet but also acts like a magnet and it does not have cubic crystals.

■ Hematite: is very weakly attracted to magnet and it may be magnetized, that is it acts like very weak magnet.

■ Franklinite: is weakly attracted to magnet.

■ Chromite: is weakly attracted to magnet.

■ Ilmenite: is weakly attracted to magnet, when heated.

■ Pyrrhotite: sometimes is strongly attracted to magnet.

■ Pyrite: sometimes very weakly attracted to magnet.

Double Refraction, if you draw a line on a paper, and put a transparent mineral on top of that line, and if you see the line as double image (two lines) through the mineral then this mineral shows double refraction. Iceland Spar (one type of calcite) shows double refraction. Look at Figure (2.16).

Taste, the mineral Halite (salt) has a salty taste.

Odor, some minerals containing sulfur, like sulfides (for example pyrite) will smell like rotten eggs if heated or struck against an object whereas sulfur (native element) will give smell that is similar to burning matches and this happens in normal room conditions. Sulfur also could give rotten egg smell if hydrochloric acid was put on it.

Some mineral containing arsenic will give garlic smell if heated or struck whereas arsenic (native element) will also give garlic smell in normal room conditions. Clay minerals (like kaolinite) have smell like fresh mud when exhaled at.

Feel, if you feel or touch some minerals, they will give you soapy feel like talc or greasy feel like graphite.

Specific Gravity, is the ratio between weight (mass) of mineral of specific volume in the air to the weight (mass) of equal volume of water displaced by the mineral (or density of object relative to density of water). Because weight (mass) divided by volume is equal to density and dividing density of mineral by density of water gives number without units therefore specific gravity has no units. Also since density of water is taken as 1 therefore, specific gravity is referred to as relative density of the mineral since specific gravity in this case is equal to density of the mineral. So for example if specific gravity of mineral is 2.5 this means the mineral weights 2.5 times as much as an equal volume of water.

To determine specific gravity for different minerals use only one hand (same hand) and when you compare two samples, they must be nearly the same size.

Specific gravity is described as heavy, medium, or light where some minerals are very heavy like native gold and native platinum and some are heavy like barite and galena. Some are or have, low weight or specific gravity like clay minerals and sulfur.

Tenacity, this character is referring to the resistance of a mineral to breaking or bending. The most important terms to describe this property are:

■ *Elastic:* the mineral layers bend under pressure, without breaking, and after you left the pressure, the layers return back to their original shape, like mica.

■ *Flexible:* the mineral will bend under pressure, without breaking, but it will not return back to its original shape, like chlorite and talc.

■ *Sectile:* Can be cut with knife like graphite and gypsum.

■ *Ductile:* Can be drawn out into strings or wires like gold due to metallic bonds.

■ *Malleable:* Can be hammered (into thin layers) or bend permanently into different shapes like gold, silver, and copper.

■ *Brittle:* Solid but breaks like glass. Diamond and quartz are some examples.

Diaphaneity, is the ability of mineral to transmit light and it is described by one of the following terms:

■ *Transparent:* is a substance that allows the light to transmit through it that is you can see through the mineral. Some samples of calcite, gypsum, quartz and halite are transparent.

■ *Translucent:* is a substance that transmits little amount of light, therefore you can't see too much through it, like some fluorite.

■ *Opaque:* substances that will not transmit light through it, therefore you can't see through it at all, like galena, pyrite, hematite, graphite, and other minerals with metallic luster.

Notice that if you make thin section of an opaque mineral, it may become translucent. And as a general rule, the thinner the slice the more it will become transparent. This general rule doesn't apply on all minerals, for example minerals of oxides and sulfides like pyrite, ilmenite, magnetite, hematite and galena will be always opaque, even in thin sections. So terms used to describe diaphaneity must be used for minerals of same thickness and the same terms could be relative when different thickness is used for different minerals (sometimes) or different thickness is used for same mineral (sometimes).

Lamellar Twinning (striations): refers to group of parallel crystals of the same mineral. For example, labradorite, mineral, that belongs to the plagioclase feldspars group. Look at Figure (2.18-7).

Play of colors: if you rotate a specific mineral under the light, you will see different colors with different angles. The mineral labradorite shows this characteristic and shows blue and violet variations. Look at Figure (2.18-7).

Reaction to acid: the mineral calcite reacts with hydrochloric acid and fizzes. Look at Figure (2.17). Another mineral that reacts with HCl acid is dolomite; however, dolomite reacts with acid only if the acid is hot or if we make powder of it then the powder will react with the acid. Sometimes dolomite takes time to react with acid (slow reaction). Also it is important to mention that, sometimes, dolomite reacts with acid directly, but not as much as calcite does (you will need magnifying glass or need to observe it carefully to see small bubbles forming) so it is easy to confuse dolomite with calcite.

Sparks of fire: if two minerals are struck together or against an object then they will produce sparks that could ignite fire. Best examples are chert and flint. Also pyrite could produce sparks if struck sharply against an object.

Figure 2.15

Figure 2.16

Figure 2.17

1: Rock crystal quartz

2: Rose quartz

3: Smoky quartz

4: Milky quartz

5: Jaspar

6: Agate

7: Labradorite plagioclase feldspar

8: Orthoclase feldspar (pink, tan)

9: Orthoclase feldspar (milky white)

10: Microcline feldspar (pink)

11: Microcline feldspar (white gray)

12: Albite plagioclase feldspar

Figure 2.18 (*Continued*)

13: Muscovite (white mica)

14: Biotite (black mica)

15: Asbestos

16: Amphibole

17: Amphibole

18: Pyroxene

19: Pyroxene

20: Olivine (light green and black)

21: Olivine (green)

22: Serpentine (dark green and green)

23: Garnet

24: Kaolinite

Figure 2.18 (*Continued*)

25: Talc (white and white pink)

26: Limonite (dark brown and black brown)

27: Hematite banded with jaspar

28: Hematite

29: Bornite

30: Bornite (specular)

31: magnetite

32: Chlorite

33: Chlorite

34: Pyrite

35: Pyrite

36: Pyrite

Figure 2.18 (*Continued*)

37: Chalcopyrite   38: Galena   39: Corundum

40: Sphalerite (regular and specular)   41: Malachite   42: Azurite

43: Calcite   44: Calcite   45: Aragonite

46: Dolomite   47: Dolomite   48: Diamond

Figure 2.18 (*Continued*)

49: Halite

50: Fluorite (purple, green, colorless, deep purple)

51: Apatite

52: Gypsum (white and pink)

53: Gypsum (satin spar)

54: Gypsum (selenite)

55: Sulfur

56: Graphite

57: Copper

58: Bauxite

59: Bauxite

60: Bauxite

Figure 2.18

PRACTICING PHYSICAL PROPERTIES OF MINERALS ACTIVITY

HARDNESS

Hardness is the ability of the mineral to resist abrasion (scratching).

So if mineral like talc, is scratched easily then this mineral will have low hardness (soft) whereas minerals that are hard to scratch have higher hardness (hard), like diamond.

Numbers are assigned to hardness so mineral of hardness (1) represents the softest mineral and mineral of hardness (10) represents hardest mineral.

Tools like fingernail (hardness = 2.5), coin (hardness = 3.5), iron nail (hardness = 4.5), glass (hardness = 5.5), and streak plate (hardness = 6.5) are used to determine the hardness of minerals. Always start with fingernail and move up to streak plate. Stop when the tool scratches the mineral sample. Use the following steps to determine the hardness of the mineral samples for questions (1)–(4). *Note* that hardness is written as range, so do not put specific number.

Is the fingernail scratching the mineral?

Yes:
You stop and the hardness of mineral is less than 2.5

No:
Use coin and see if coin is scratching the mineral:

No:
Use glass and see if glass is scratching the mineral:

Yes:
You stop and the hardness of the mineral is between 2.5 and 3.5

No:
Use streak plate and see if streak plate is scratching the mineral:

Yes:
You stop and the hardness of the mineral is between 3.5 and 5.5

No:
Then the mineral hardness is more than 6.5
There are no other tools (in your set) used to determine hardness of minerals harder than 6.5

Yes:
You stop and the hardness of the mineral is between 5.5 and 6.5

1) The hardness of sample (4) =

2) The hardness of sample (29) =

3) The hardness of sample (13) =

4) The hardness of sample (26) =

CLEAVAGE

Is the ability of the mineral to break along planes of weakness. Cleavage planes are usually flat and shiny, and sometimes they look like steps.

Some minerals have cleavage (planes or directions) and some do not have cleavage (planes or directions) and sometimes the shiny flat surfaces are not cleavage planes but crystal faces.

You know if a mineral will have cleavage or not by practice and this takes time.

Cleavage is described by number of planes (directions) and the angles between two adjacent planes. Look carefully at Figures (2.5)–(2.10) and study the different types of cleavage and the differences between them.

Now put the following samples in front of you.

Samples: 3, 4, 11, 12 (or 12a), 13, 14, 15, 22, and 27.

Try to match the description given for different types of cleavage to the number of sample that best matches that specific description:

5) This mineral has one direction (plane) and these planes are parallel to each other like sheets of book. This cleavage is called *basal cleavage*.

   Sample # (   ) and sample # (   ) show this type of cleavage.

6) Find clear or nearly clear sample that looks like a cube (sometimes you have broken cubes). Note that there are three directions (planes) and the angles between cleavage planes is nearly 90°.

   This cleavage is called *isometric or cubic cleavage*.

   Sample # (   ) shows this type of cleavage.

   Try to find another sample that shows isometric cleavage (the sample is gray and metallic):

   Sample # (   ) shows this type of cleavage.

7) Find another clear or nearly clear sample (sometimes white) that have three directions (planes) and the angles between cleavage planes *is not* 90°. This is called *Rhombohedral cleavage*.

   Sample # (   ) shows this type of cleavage.

8) This type of cleavage is more difficult to see and it shows two planes (directions) that are 90° to each other. This cleavage is called *Prismatic cleavage*.

   Sample # (   ) and sample # (   ) show this type of cleavage.

9) This type of cleavage shows four directions (planes) of cleavage that are not 90° to each other.

   This cleavage is called *Octahedral cleavage*.

   Note that some samples are small and are complete (perfect) whereas some samples are big and broken. Samples could show different colors.

   Sample(s) # (   ) shows this type of cleavage.

10) This type of cleavage shows six directions (planes) of cleavage that are not 90° to each other.

   Sample # (   ) shows this type of cleavage.

   Note that it is difficult to see all the cleavage planes since some samples are broken or not complete.

FRACTURE

Put samples (9), not found in all boxes, (2), (23) , and (32) in front of you and notice that they do not show flat surfaces but random surfaces that are not shiny. These are called fractures.

Fractures are described by terms like conchoidal, hackly, uneven (irregular) and even (regular).

11) Sample (9)—not found in all boxes – shows _____ fracture.

12) Samples (2), (23), and (32) show _____ fracture.

STREAK

Bring the white streak plate and rub a mineral on the streak plate. You will notice powder on the streak plate. That powder could be white, or any other color. This is called streak, so streak is the color of the powder, the mineral leaves on the streak plate.

Notice that hardness of streak plate is 6.5 so any mineral harder than streak plate will scratch it and therefore *will not* show the streak of the mineral and in this case the mineral will show no streak.

Put the following minerals in front of you then answer the following questions.

Samples: 16 (or 16a), 20, 24, 35, and 37.

Try to match the samples numbers with the description of streak and mineral.

13) Soft mineral with black gray streak: Sample # (   ).

14) Brass yellow mineral with greenish black streak: Sample # (   ).

15) This mineral shows brownish red streak: Sample # (   ).

16) An olive yellowish green to blackish green mineral with white streak: Sample # (   ).

17) Blackish to dark gray mineral that sometimes look like sample in question (15) but with black to dark gray streak: Sample # (   ).

18) Corundum (sample # 90)—not found in box—has hardness of (9). Corundum shows (white streak, pale dark streak, no streak). Select the correct answer. *Hint:* you do not need to do the streak test.

19) Put sample (11) and sample (12) or (12a) in front of you. Notice the color of the two samples. These two samples are two different types of the same mineral and both have same hardness of 6. Determine the streak of the two samples (if it is possible):

Sample (11) shows _____ streak.

Sample (12) or (12a) shows _____ streak.

After reading the above and knowing the colors of the two samples and their streak, is it true that different types of the same mineral always show the same streak?

LUSTER

Luster means, how does the surface look like when it reflects light, or simply, the appearance of a mineral surface in a reflected light.

There are two types of luster:

1. Metallic: the minerals have metal surfaces like iron, gold and copper, which are shiny. Note that "shiny" does not mean the mineral is metallic, so not every shiny mineral will have metallic luster. Usually these minerals are heavy, opaque, and give dark streak.

   Examples of minerals with metallic luster are: galena, pyrite, graphite, and some hematite.

2. Nonmetallic: the minerals don't have metal appearance and look like glass or colored glass and are usually light colored, transparent to translucent and light in weight. The streak is usually colorless to very light color. Dull (not shiny) minerals are also considered nonmetallic. Other terms are used to further describe the nonmetallic luster like vitreous, silky, adamantine, and so on.

Answer the following questions by selecting metallic (M) or nonmetallic (NM) luster:

20) Sample # (3) shows _____ luster.

21) Sample # (4) shows _____ luster.

22) Sample # (14) shows _____ luster.

23) Sample # (18) shows _____ luster.

24) Sample # (23) shows _____ luster.

25) Sample # (24) shows _____ luster.

26) Sample # (26) shows _____ luster.

27) Sample # (30) shows _____ luster.

28) Sample # (35) shows _____ luster.

29) Sample # (37) shows _____ luster.

SPECIFIC GRAVITY OR RELATIVE DENSITY

Is simply referring to how heavy is the mineral. So minerals could be heavy, medium, or light.

To determine the specific gravity in a very simple way simply "heft" the sample using one hand only. Do not use two hands to weight two samples. Always weight two samples with the same hand. Also note, that samples that are compared should be nearly the same size so when you weigh them, take into consideration the size of the two compared samples.

Put the following samples in front of you: 5, 14, 20, and 23.

Try to match the correct name of mineral to its correct specific gravity/relative density:

30) Sample # (   ) is sulfur. Sulfur is the lightest.

31) Sample # (   ) is olivine. Olivine is lighter than magnetite.

32) Sample # (   ) is magnetite. Magnetite is medium.

33) Sample # (   ) is galena. Galena is the heaviest.

OTHER PROPERTIES

Some of the minerals are identified by their specific characteristic.

34) Some minerals have specific smell, like sulfur that gives rotten egg smell when you add hydrochloric acid to it and some of the minerals smell like fresh mud when you exhale on it like kaolinite.

Put sample (5) and sample (32) in front of you and using the above information try to name the following samples:

Sample (5) is:

Sample (32) is:

35) Put samples (11) and (12) or (12a) in front of you and notice that one sample out of the two shows either striations (sometimes known as Lamellar Twinning) and these are parallel lines or it shows play of colors (change of colors from blue to purple when mineral is rotated) like the mineral labradorite (plagioclase feldspar). Labradorite is sometimes blue to gray white with some blue stains. The other sample shows pink color (like flesh) or creamy white color like the mineral orthoclase (potassium feldspar).

From the above information:

Sample (11) is:

Sample (12) or (12a) is:

36) Some minerals react with hydrochloric acid that is they fizz. This property is known as acid reaction (try to see the bubbles using magnifying glass or try to hear the sound of fizzing by putting the sample close to your ear). Best example that shows this property is the mineral calcite.

Some minerals are attracted to magnet and this is known as magnetism. Best example is magnetite.

Some minerals are salty (have taste) like halite. Some minerals show elastic layers like the mineral biotite and muscovite (both are two types of mica).

Put the following samples in front of you and based on the information given above try to identify (name) the following samples:

The samples: (3), (4), (15), (20), and (22).

Sample (3) is:

Sample (4) is:

Sample (15) is:

Sample (20) is:

Sample (22) is:

37) *Diaphaneity*

Refers to the ability of mineral to transmit light, so minerals are transparent (allow all or most of light to pass through), translucent (allow some of light to pass through) or opaque (no light pass through).

Put samples (11), (15), and (36) in front of you and see if they are transparent, translucent, or opaque.

Sample (11) is:

Sample (15) is:

Sample (36) is:

38) *Tenacity*

Is the ability of minerals to resist bending or deformation. Some terms used to describe tenacity include the following: elastic, flexible, sectile, ductile, malleable, and brittle.

Look and examine the following samples:

Samples (3) and (22) are two types of mica. Mica shows elastic sheets.

Samples (35), (18), and (18a) are sectile.

Diamond (not in the box) and glass (obsidian) are brittle.

Table (A)

Metallic to Submetallic Luster

Name/ Chemical composition	Hardness	Color	Streak	Cleavage Directions and angle	Other Properties
Graphite C	1 – 2 Softer than fingernail	Gray, black gray	Gray to black	Yes (Prominent) Sometimes not seen or not clear	Greasy feel; flexible plates.
				1 dir. (platy)	
Galena PbS	2.5 Softer than glass but harder than fingernail	Gray	Gray	Yes (Prominent)	Heavy.
				3 dir., 90°	
Copper Cu	2.5 – 3 Softer than glass but harder than fingernail	Brown red with tarnish	Red brown	No (Absent)	Showa hackly fracture; ductile.

Name/ Chemical composition	Hardness	Color	Streak	Cleavage		Other Properties
				Directions and angle		
Bornite Cu_5FeS_4	3 Softer than glass but harder than fingernail	Bronze but with copper-red to purplish iridescent	Grayish black	No (Absent)		Known as peacock ore.
Chalcopyrite $CuFeS_2$	3.5 – 4 Softer than glass but harder than fingernail	Brass yellow, tarnish usually	Greenish black to oily black	No (Absent)		Looks like pyrite but the hardness is less than pyrite.
Limonite $FeO(OH)H_2O$	4 – 5.5 Softer than glass but harder than fingernail and sometimes softer than fingernail	Yellow-brown, Orange-brown, Dark brown	Yellowish brown to reddish	No (Absent)		
Goethite $(Fe)(O)(OH)$	5 – 5.5	Very dark brown to dark gray	Brownish yellow	No (Absent)		
Hematite Fe_2O_3	5.5 – 6.5 Sometimes < 2 Harder than glass and sometimes softer than fingernail	Silver gray to reddish brown	Reddish brown	No (Absent)		Sometimes hematite is nonmetallic. Hardness varies depending on oxidation.
Magnetite Fe_3O_4	5.5 – 6.5 Harder than glass	Black	Black	No (Absent)		Attracted to magnets.
Pyrite FeS_2	6 – 6.5 Harder than glass	Brass yellow to golden yellow	Greenish to brownish black to oily black	No (Absent)		Shows crystal forms but also massive and granular.

Table (B)

Nonmetallic Luster

Name/ Chemical composition	Hardness	Color	Streak	Cleavage	Luster	Other Properties
				Directions and angle		
Talc $Mg_3Si_4O_{10}(OH)_2$	1 Softer than fingernail	Light green to white gray	White	Yes (Prominent) Sometimes not seen or not clear	Pearly or greasy	Called soapstone; soft; have flexible sheets.
				1 dir., basal		

(Continued)

Name/ Chemical composition	Hardness	Color	Streak	Cleavage		Luster	Other Properties
				Directions and angle			
Sulfur S	1.5 – 2.5 Softer than fingernail	Yellow	White	No (Absent)		Resinous	Smells like rotten egg when HCl acid applied.
Realgar AsS	1.5 – 2 Softer than fingernail	Red to orange	Orange red to orange yellow	Yes (Prominent) 1 dir		Resinous to greasy	Sectile; poisonous; if exposed to light for long period it will disintegrate to orange yellow powder.
Gypsum $CaSO_4\ 2H_2O$	2 Softer than fingernail	Colorless, white, gray yellowish, pink	White	Obvious (prominent) in selenite only and it is 3 dir., $\neq 90°$ and no cleavage (absent) in other types		Vitreous, pearly	Have three varities; 1) Selenite, which is clear and transparent. 2) Satin spar, fibrous. 3) Massive gypsum (alabaster), usually granular or structureless.
Kaolinite (One type of clay minerals) $Al_2Si_2O_5(OH)_4$	2 – 2.5 Softer than fingernail	White to gray	White	Yes (Prominent) Sometimes not seen or not clear 1 dir., basal		Dull, earthy	
Halite NaCl	2.5 Softer than glass but harder than fingernail	Colorless, yellow	White	Yes (Prominent) 3 dir., 90°		Vitreous	Salty taste.
Biotite (One type of mica called black mica) $K(Mg,Fe)_3(AlSi_3O_{10})(OH)_2$	2.5 Softer than glass but harder than fingernail	Black brown	Gray to white	Yes (Prominent) 1 dir., basal		Vitreous	Elastic sheets.
Muscovite (One type of mica called white mica) $KAl_2(AlSi_3O_{10})(OH)_2$	2.5 Softer than glass but harder than fingernail	Colorless, gray or light green	White	Yes (Prominent) 1 dir., basal		Vitreous	Elastic sheets.

Name/ Chemical composition	Hardness	Color	Streak	Cleavage		Luster	Other Properties
				Directions and angle			
Chlorite $(Mg,Fe,Al)_6SiAl_4O_{10}(OH)_8$	2 – 3	Green but also yellow, brown	Green	Yes (Prominent) Sometimes not seen or not clear		Vitreous	Flexible sheets.
				1 dir., basal			
Serpentine $Mg_3Si_2O_5(OH)_4$	2.5 – 4	Shades of green but also brownish gray, white or yellow	White	Shows basal cleavage (prominent) in antigorite but none (absent) in fibrous chrysotile		Waxy, greasy, silky	
Bauxite Mixture of: AlO(OH), Al(OH) and HAlO$_2$	2 – 7	White to brown	White to light brown	No (Absent)		Earthy/dull	Mixture of three clay minerals; boehmite, gibbsite, diaspore. Earthy odor when breathed on.
Calcite $CaCO_3$	3 Softer than glass but harder than fingernail	Colorless, white, white yellow	White	Yes (Prominent)		Vitreous	The clear transparent calcite is icelandspar. Reacts with HCl acid. Sometimes shows double refraction.
				3 dir., ≠ 90°			
Dolomite $CaMg(CO_3)_2$	3.5 – 4.5 Softer than glass but harder than fingernail	White, gray, brown, pink	White	Not obvious (absent) in rock masses or when massive. However it shows 3 dir., ≠ 90° (prominent)		Vitreous to pearly	Reacts with HCl acid when acid is warm or when powdered. Dolomite sometimes massive and sometimes crystalline.
Malachite $Cu_2(CO_3)(OH)_2$	3.5 – 4 Softer than glass but harder than fingernail	Bright green	Pale green	Not obvious (absent) in massive forms		Dull	Sometimes shows banding; associated with azurite.

(Continued)

Name/ Chemical composition	Hardness	Color	Streak	Cleavage		Luster	Other Properties
				Directions and angle			
Azurite $Cu_3(CO_3)_2(OH)_2$	3.5 – 4 Softer than glass but harder than fingernail	Shades of deep blue	Light blue	No (Absent)		Dull to vitreous	Associated with malachite.
Sphalerite ZnS	3.5 – 4 Softer than glass but harder than fingernail	Dark brown or black to yellow	Yellow to brown	Yes (Prominent)		Resinous	
				6 directions ≠ 90°			
Fluorite CaF_2	4 Softer than glass but harder than fingernail	Colorless, yellow, purple, green	White	Yes (Prominent)		Vitreous	
				4 directions ≠ 90°			
Apatite $Ca_5(PO_4)_3(F,Cl,OH)$	5 Softer than glass but harder than fingernail	Shades of green, yellow, blue, brown	White	Basal (1 dir.) (prominent) although sometimes not seen or not clear		Vitreous	
Limonite $FeO(OH)H_2O$	4 – 5.5 Softer than glass but harder than fingernail and sometimes softer than fingernail	Yellow-brown, Orange-brown, Dark brown	Yellowish brown to reddish	No (Absent)		Earthy, dull	
Goethite $(Fe)(O)(OH)$	5 – 5.5	Very dark brown to dark gray	Brownish yellow	No (Absent)		Dull, although sometimes adamantine	
Hematite Fe_2O_3	5.5 – 6.5 Sometimes < 2 Harder than glass and sometimes softer than fingernail	Red to reddish brown	Reddish brown	No (Absent)		Earthy, dull	Hematite could be specular (metallic), massive dull and also oolitic.
Augite (One type of Pyroxene) Complex silicate	6 Harder than glass	Dark green to blackish green	Light gray	Yes (Prominent)		Vitreous	Similar to hornblende but crystals are blocky.
				2 dir., = 90°			

Name/ Chemical composition	Hardness	Color	Streak	Cleavage		Luster	Other Properties
				Directions and angle			
Hornblende (One type of Amphibloe) Complex silicate	6 Harder than glass	Black to dark green	Grayish white	Yes (Prominent) 2 dir., ≠ 90°		Vitreous	Similar to augite but crystals are prismatic.
Olivine $(Mg,Fe)_2SiO_4$	6 Harder than glass	Olive green to yellow green	White or gray	No (Absent)		Vitreous	Usually is glassy and granular, and sometimes it shows sugary masses of yellowish green color.
Labradorite (One type of Plagioclase Feldspar)	6 Harder than glass	Dark gray to blue	White	Yes (Prominent) 2 dir., = 90°		Vitreous	Sometimes shows lamellae twinning or blue play of colors.
Microcline/Orthoclase (One type of Potassium feldspar)	6 Harder than glass	Creamy white, tan to pink orange, green	White	Yes (Prominent) 2 dir., = 90°		Vitreous	Shows perthitic texture. Green color variety of potassium feldspar is known as amazonite.
Quartz (Crystalline variety) SiO_2	7 Harder than glass	Colorless, white, gray, pink, black, purple, yellow, green	White	No (Absent)		Vitreous	Varieties: rock crystal, milky, smoky, rose, amethyst, citrine.
Quartz (Chalcedony variety) SiO_2	7 Harder than glass	All colors observed	White	No (Absent)		Waxy or dull	Varieties: -Agate; banded -Jaspar; red or brown -Chert or Flint; gray, white or black and if rubbed together they give smoke.

(*Continued*)

Name/ Chemical composition	Hardness	Color	Streak	Cleavage		Luster	Other Properties
				Directions and angle			
Garnet Silicates of Al, Ca, Mg, Fe, Mn, Cr	6.5 – 7.5 Harder than glass	Red, brown and also yellow, pink, green, black	White	No (Absent)		Vitreous or resinous	Shows parting.
Corundum Al_2O_3	9 Harder than glass	Gray, brown, blue, red, colorless	White	No (Absent)		Vitreous, dull, adamantine	Shows parting; barrel shaped crystals; gem varieties as sapphire (blue), ruby (red).
Diamond (C)	10 Harder than glass	Colorless	White	No (Absent)		Adamantine	Could develop conchoidal fractures. Brittle and has a cold feel.

Mineral Number	Color	Luster	Hardness	Cleavage	Streak	Other Properties	Mineral Name
1							
2							Bauxite
3							
4							
6							
11							
12							
13							
14							
15							
16							
18							
18A							
19							
19A							
20							
21							
22							
23							
24							
25							Augite
26							
27							Sphalerite
29							
30							
31							
32							
33							

(Continued)

Mineral Number	Color	Luster	Hardness	Cleavage	Streak	Other Properties	Mineral Name
34							Chlorite
35							
36							Augite
37							Serpentine

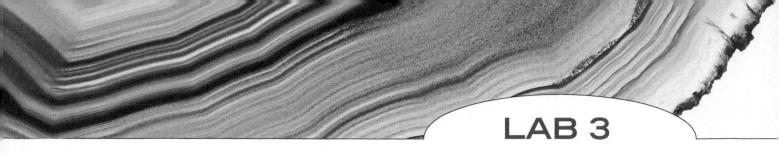

SEDIMENTARY ROCKS IDENTIFICATION

Try to follow the steps <u>in order</u> and <u>do not jump</u> to other steps unless you make sure that you understand the information and answer the questions in each step.

STEP (1): SEDIMENTARY STRUCTURES

Before identifying sedimentary rocks, it is important to know about sedimentary structures. Sedimentary structures include fossils, ripple marks, mud cracks, horizontal layers, and cross bedding. Look at samples # 64, 71, 74, and 83. Do you find any sedimentary structures in these samples? To answer this question, look at sedimentary structures shown in Figure (3.1) then fill Table (3.1).

Note that some samples do not show any sedimentary structure(s) and some samples may show more than one sedimentary structure.

Table 3.1

Sample Number	Type of Sedimentary Structure(s) (If Found)
64	
71	
74	
83	

STEP (2): CLASTIC/DETRITAL SEDIMENTARY ROCKS

Put samples # 53, 56, 70, 71, 74, and 77 in front of you. Now use a ruler supplied by your instructor (in millimeter, mm) to determine the grain size for each sample. Remember that 1 mm is equal to the thickness of one individual hair sample. If the ruler did not help or is confusing then use sand grains as your reference tool with Figure (3.2) to help you. Use the following as a guideline to determine the grain size and then fill column (2) of Table (3.2):

- If the general particles composing the rock are larger than sand grains then particles are coarse (more than 2 mm).

- If the general particles composing the rock are same size as sand grains then particles are medium (between 2 and 1/16 mm).

- If the general particles composing the rock are smaller than sand grains then particles are fine (between 1/16 and 1/256 mm for silt and less than 1/256 mm for clay).

Fill the answers in Table (3.2), column (2) only.

Sedimentary Structures

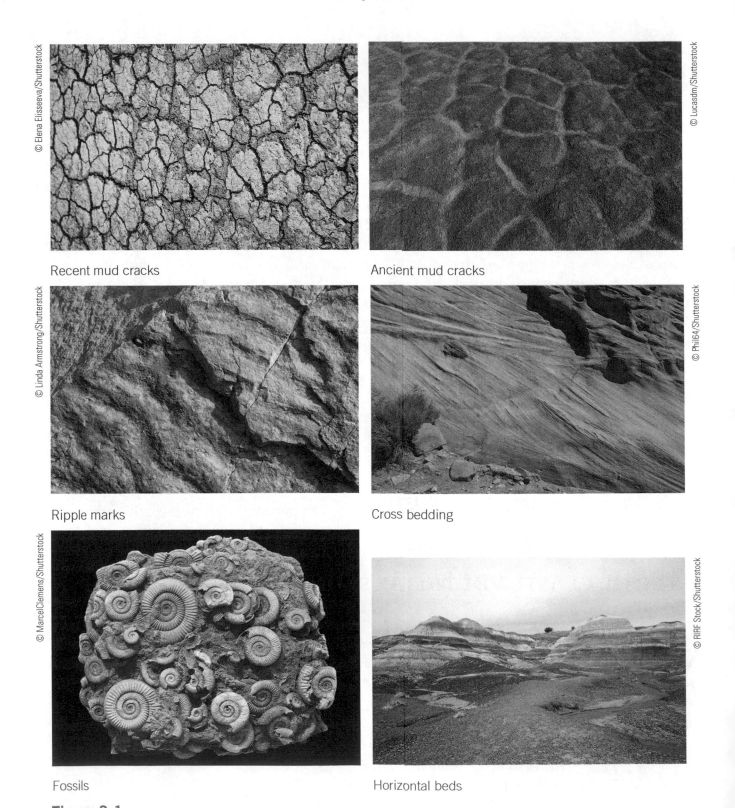

Recent mud cracks

Ancient mud cracks

Ripple marks

Cross bedding

Fossils

Horizontal beds

Figure 3.1

Gravel

Gravel

Sand (sorting is 90%)

Green to brown sand (sorting is 100%)

White sand (sorting is 100%)

Silt/clay

Figure 3.2

Figure 3.3

Figure 3.4

(2.1): Based on column (2) of Table (3.2), and using the information given to you below, fill column (3) of Table (3.2):

* If particles are >2 mm then the grains are coarse grains.

* If particles are between 2 and 1/16 mm then the grains are medium grains.

* If particles are <1/16 mm then the grains are fine grains.

Table 3.2

Column (1)	Column (2)	Column (3)	Column (4)
Sample Number	Grain Size (mm) (Write the Value)	Grain Size in General (Coarse, Medium, Fine)	Name of Rock
53			
56			
71			
77			
70			Siltstone
74			

(2.2): Before you fill column (4) of Table (3.2), answer the following questions, since they will help you to determine the name of the rocks with coarse grains:

(A) Put all the samples (from the above list) that have only *large grains* (coarse grains) together, in front of you, then compare the shape of the grains:

   (A1) Do the shape of grains in sample # 53 look like those in Figure (3.3) or those in Figure (3.4)?

   (A2) Do the shape of grains in sample # 56 look like those in Figure (3.3) or those in Figure (3.4)?

   After knowing the answers for questions (A1) and (A2) and using Table (3.3) put your answers in column (4) of Table (3.2).

(B) Now put all the samples that have *medium grains* together, in front of you. Medium-grained rocks usually have sugary feeling. Feel those rocks with your hand.

   These sugary feeling rocks are usually different types of sandstone. Using Figures (3.5) and the information given to you below, try to name the different types of sandstone and then fill column (4) of Table (3.2).

   • Red sandstone: made of quartz and the red color is due to either iron oxide as cement or the existence of feldspars minerals.

   • Quartz sandstone: mature type of sandstone, white and made of 75% quartz. Well sorted with rounded grains however some quartz sandstones are red to brown to orange due to iron oxide as cement or stain. The cement in quartz sandstone could be white calcite or quartz.

   • Banded sandstone: different colored bands of white quartz and red iron oxide band due to different characteristics of sand in each layer due to different depositional (environment) characteristics. Banding could result from different mineral (chemical) composition or from difference in mineral's grain size.

   • Arkose: sandstone that is more mature than greywacke and is red to brown and feels sugary with some larger grains. Made of quartz and more than 25% feldspars. Poorly sorted with more angular (less rounded) fragments. The red to brown and sometimes yellowish color is due to iron oxide as stain or cement. May contain calcite as cement and if grains are coarse then it could be considered conglomerate.

- Graywacke sandstone: immature sandstone with dark color and is made of quartz and other minerals like feldspar, clay, mica, and rock fragments, and rock fragments are angular and poorly sorted. Arkose is sometimes called "dirty sandstone."

(C) Put all the samples that have *fine grains* together, in front of you. There may be one or two samples only.

Sedimentary rocks with fine grains are those rocks that are smooth (not sugary).

Siltstone is one example. Shale is another example. The grains of siltstone and shale are too small to be seen by naked eyes and you need magnifying lenses or even microscope to see the grains and hands or teeth can be used to differentiate between silt and clay particles. Silt has gritty grain feeling whereas clay has smooth feeling like chalk dust or powder.

Shale is either black (rich in organic material) called black shale or gray green (rich in clay minerals which are also clay sized particles) called gray green shale. Shale crumbles easily with pressure because shale lacks cement. In addition, shale is made of layers (fissile, that is, it splits in to platy slabs parallel to bedding). Mudstone is same as shale; however, it does not show layers but it breaks as blocks or chunky pieces.

Which one of the two samples (70) and (74) is shale? Fill column (4) of Table (3.2).

Note: Sometimes, it is difficult to differentiate between shale and siltstone due to the fine texture therefore siltstone is given as the sample number (70).

Table 3.3

Grain Size	Grain Size (mm)	Name of Sediments	Name of Rock	Minerals	Characteristics
Coarse	>2	Gravel	Conglomerate Breccia	Rock fragments and quartz	Conglomerate has rounded fragments whereas breccia has angular fragments
Medium	2–1/16	Sand	Sandstone Arkose	Quartz Quartz + Feldspars	Sandstone has sugary feeling and sometimes the color is white, yellow red to red. Arkose is poorly sorted and has angular fragments with reddish brown color.
Fine	1/16–1/256	Silt	Siltstone	Clay (also quartz and feldspars)	White to yellow white
Very fine	<1/256	Clay	Shale	Clay (with very little amount of quartz)	Black to dark gray or gray green and shows layering. Smooth flat and structureless. Sometimes shows fossils. Smells like fresh mud and may react to acid due to existence of calcite as cement

STEP (3): BIOCHEMICAL/CHEMICAL SEDIMENTARY ROCKS

Put samples # 58, 64, 78, 83, 84, and 57 in front of you to study them.

Bring HCl (hydrochloric) acid and put two drops on each sample. If any of the rocks reacts with acid (fizzes) then it is either limestone or dolomite. Use magnifying glass to see the bubbles.

For those samples *that did not react with acid*, put them on side and then apply (3.1) on page (45). Do this step after identifying dolomite and the different types of limestone.

Now those samples that fizzed with acid are either limestone or dolomite and to differentiate between them, notice the following observations:

Limestone fizzes easily with HCl acid and you can see the bubbles with your naked eye whereas dolomite will fizz less than limestone and only in the following conditions:

1. If HCl acid is warm.
2. If you make powder of dolomite and the powder will fizz with HCl.
3. If time is given for dolomite to fizz that is you need to wait for few seconds to see very small bubbles forming. If you are not able to see the bubbles then use magnifying glass or hold the sample close to your ears to hear the "fizzing."

Note: If you do not have the HCl acid, try to use Table (3.4) to differentiate between dolomite and the different types of limestone. Also use Figures (3.5) to aid you in the identification.

It is important to say that when using Table (3.4) you need to be careful and try to observe the features and at the same time compare between them to notice the differences. Sometimes Table (3.4) works and sometimes it will not work and this depends on the samples.

After knowing which samples is limestone and which sample is dolomite, it is important to recognize the different types of limestone by observing the following features in Table (3.4):

Table 3.4

Limestone	Made of 100% of shell fragments or fossils. Poorly cemented.	Coquina	
	Made of some shell fragments or fossils (skeletal). Cemented with calcite. Usually white but sometimes gray black. More dense than coquina.	Fossil limestone	
	Sometimes it is difficult to differentiate between them. Both may be smooth and massive (structureless), light white to yellowish or brownish. The dark gray to black color is due to organic material. Crystalline limestone shows more crystals seen by naked eye than micrite (very small crystals that are difficult to see by naked eye).	Micrite/crystalline limestone	
	Oolitic refers to the very small spheres that look like the letter "O" and is cemented by calcite. The spheroidal particles are usually equal or less than 2 mm in diameter.	Oolitic limestone	
	Massive and structureless. White to light gray or light green and more irregular than micrite. Soft. Poorly cemented. Made of microscopic shells of foraminifera. Leaves powdery marks on surfaces.	Chalk (Not in box)	
Dolomite	Smooth and massive (structureless). More irregular than micrite/crystalline limestone. White and sometimes show yellowish color.	Dolomite	

After you finish, write the names of dolomite and the different types of limestone in Table (3.5)

Table 3.5

Sample Number	Reaction to Acid - Only for Biochemical/Chemical Type (No reaction, Weak Reaction, Strong Reaction)	Name of Rock
58		
64		
78		
83		
84		
57		
Organic Sedimentary Rocks		
Sample Number	Acid test	**Type of Coal**
55	Not applicable	

(3.1): Chert and Flint: Will not react with HCl acid, shows smooth surfaces, structureless, has variations of colors from white, brown to gray black and shows no layering. If rubbed with another flint or chert then it will give "smoky" smell or sparks.

Chert or flint are chemical/biochemical sedimentary rocks and are made of the mineral quartz.

If chert or flint reacted with HCl acid then this is because some of chert and flint form from silica replacement of limestone.

Write the name of the sample in table (3.5).

STEP (4): ORGANIC SEDIMENTARY ROCKS

The final group to study is the organic sedimentary rocks represented by coal.

Coal has three ranks described as follows in Table (3.6):

Table 3.6

Peat (Low Rank)	Lignite (Medium Rank)-Dirty Coal	Bituminous (High Rank)
Brown to dark brown	Dark brown to black	Black
Soft (very brittle), crumbles	Hard (brittle), leaves dusty soil	Harder (brittle), layered
Shows plants remains	May show plants (fossil) remains	No plants remains
Dull	Not dull but not shiny	More shiny

Put sample (55) in front of you and then based on the information mentioned in Table (3.6) decide if that sample of coal is peat, lignite or bituminous. Write your answer in Table (3.5).

1: Conglomerate

2: Breccia

3: Red sandstone

4: Banded red sandstone

5: White sandstone (quartz sandstone)

6: Brown-red arkose

Figure 3.5 (*Continued*)

7: Pink-red arkose

8: Siltstone

9: Black shale

10: Black shale

11: Fossil shale

12: Gray shale

Figure 3.5 (*Continued*)

13: Coquina

14: Fossil limestone

15: Oolitic limestone

16: Oolitic limestone (black)

17: Crystalline limestone

18: Crystalline limestone (gray brown)

Figure 3.5 (*Continued*)

19: Dolomite

20: Dolomite

21: Dolomite

22: Dolomite

23: Dolomite

24: Chert

Figure 3.5 (*Continued*)

25: Chert

26: Flint

27: Flint

28: Peat coal

© Swapan Photography/Shutterstock, Inc.

29: Bitaminuos coal

Figure 3.5

IGNEOUS ROCKS IDENTIFICATION

To classify igneous rocks, you need first to identify the rock's texture (rate of cooling) and the rock's mineral composition (reflected on color) then use the Igneous Rocks Identification Table to name the rock. Read the guidelines, understand them, and then follow them in order.

1. **Rock's Texture:**

   Rock's texture reflects rate of cooling and therefore it is an indication of grain size.
   Take the rock sample and look at the grain size of the rock and follow the guidelines:

   Note: to measure the grain size, either use a ruler or use the scale in Figure (4.1). The scale uses both centimeters (millimeters subdivisions) and inches units.

   1A)   – If the grains are between 2 and 10 mm in diameter (are visible by naked eyes and all crystals have nearly the same size) or

           – If you can see all the minerals, represented by different colors and these different colors of minerals are covering the *whole* rock or

           – If you can see shiny crystals, that covers the *whole* rock

           Then the rock has phaneritic texture. Look at Figure (4.3-2, 3, 4, 9, 10, 14).

           Rocks with phaneritic texture are considered intrusive rocks, and this type of texture is produced by slow rate of cooling.

   1B)   – If the grains are less than 2 mm in diameter (small crystals that cannot be seen by naked eye but using hand lens) or

           – If you can see *some* minerals, not covering the whole rock or

           – If you can see *some* shiny crystals, not covering the whole rock

           Then the rock has aphanitic texture. Look at Figure (4.3-5, 6, 11, 12, 13, 15).

Fill th

Rocks with aphanitic texture are considered extrusive, and this type of texture is produced by fast rate of cooling.

1C)   If the rock shows vesicles, voids, holes, or openings like sponge then the texture is cellular. The term frothy, vesicular is applied to rocks with small and sharp-edged vesicles. Look at Figure (4.3-16, 17, 18).

Rocks with cellular (vesicular) texture are considered extrusive, and this type of texture is produced by very fast rate of cooling.

Note that these vesicles are not minerals.

If some rocks contain holes and the amount of holes are less than the mass of the solid rock then it is ok to put the word "vesicular" in front of the name of rock like vesicular basalt. However, if the amount of holes are more than the mass of the solid rock then the rock is called either "pumice" or "scoria." Furthermore, if pores or vesicles are filled with secondary minerals after the lava solidifies then the rock will show amygdaloidal texture.

1D)   If the rock shows smooth surface (no minerals or crystals or vesicles) like glass then the rock has glassy texture. Look at Figure (4.3-19, 20).

Rocks with glassy texture are considered extrusive and this type of texture is produced by very fast rate of cooling.

1E)   If the rock shows two different sizes of grains (crystals) then the rock has porphyritic texture. This type of texture should show two distinctive different sizes of crystals and not gradual sizes of crystals.

The large crystals are called phenocrysts, and the small crystals are the groundmass or matrix and this texture indicates two rates of cooling; a slow rate of cooling that gave large crystals followed by fast rate of cooling that gave small crystals surrounding the large crystals. Look at Figure (4.3-2, 5, 6, 11, 12, 13).

The porphyritic texture could develop in plutonic (intrusive) rocks and also could develop in volcanic (extrusive) rocks although this type of texture is more common in volcanic rocks.

In plutonic rocks, the phynocrysts are surrounded by a coarser matrix and the texture is called porphyritic phaneritic (phynocrysts occur within phaneritic matrix), whereas in volcanic rocks, the phynocrysts are surrounded by a finer matrix, and the texture is called porphyritic aphanitic (phynocrysts occur within aphanitic matrix).

1F)   If the rock is made of rock fragments of different sizes (usually not minerals) produced by volcanic eruptions then we refer to this type of texture as pyroclastic texture; therefore, this type of texture refers to large rock fragments in volcanic ash.

Although this type of texture could be found in acidic/felsic, intermediate, and basic/mafic igneous rocks, it is more common in acidic/felsic and intermediate igneous rocks since the magma that produces these rocks are more viscous and contain more gases.

Pyroclastic texture indicates that the rock is extrusive. Look at Figure (4.3-24).

1G)   If the rock is made of crystals that are larger than 10 mm then the rock has pegmatitic texture. This type of texture develops not only because of the slow rate of cooling but also because of existence of water that helps the migration of chemical elements to the growing crystals. Rocks with pegmatitic texture are sometimes rich in rare minerals like beryl and tourmaline. Look at Figure (4.3-1).

2. **Rock's Color:**

Color is an indication of mineral composition. Color gives the name of the group of igneous rock, that is, if the rock is acidic/felsic, intermediate, or basic/mafic.
Take the rock sample and look at the color of the rock and follow the guidelines:

2A)   The rock is acidic or felsic if:

      2A-1) the general color of the whole rock is pink or shades of pink (salmon, peach, tan, orange, or close to red). Look at Figure (4.3-5, 6).

      2A-2) some pink-colored crystals or minerals are present among other colored minerals (black, gray, white). Look at Figure (4.3-1, 2, 3, 4).

      2A-3) the color of the rock is light gray. This may be confusing because gray may show different shades of darker or lighter gray. Always try to compare two samples that have gray colors and see which one is lighter than the other one. Look at Figure (4.3-7, 8).

2B)   The rock is intermediate if the color is gray (not light gray and not dark gray), gray green, light green with light purple or is combination of equal amount (percent) of black (dark) and white (light) minerals.

Look at Figure (4.3-9, 10, 11, 12, 13).

2C)   The rock is mafic/basic if:

      2C-1) the color is black and the texture is *not* glassy.
                Look at Figure (4.3-14, 15, 18).

      2C-2) the color is dark gray. Try to compare two samples that have gray colors and see which one is darker than the other one.

      2C-3) the color is dark green.

2D)   If the color of the sample is black and the texture *is* glassy then you have to do the test in order to decide whether the rock is acidic/felsic or basic/mafic. Therefore, the following test is done only for obsidian.

The Test: Hold the sample in front of a light source and see if light is passing through the sample or at least passing through the edges of the sample. If you see the light through the sample (even small portion) then the sample is acidic/felsic. If light is not passing through the sample at all, then the sample is basic/mafic.

Remember that you do this test *only* for those samples that have black color and show glassy texture (obsidian).

If it is difficult to recognize the color of the rock or if the color guidelines are confusing then use Igneous Rocks Color Index Tables (Figure 4.2). To use this color index simply compare the color of the rock sample to the colors of index (A), index (B), and index (C). The last strip of index (C) is comparison based on % of *dark* minerals. For example, if dark minerals make more than 45% of the total percent of minerals then it is possible that this rock is basic/mafic igneous rock.

3. After knowing the texture of the rock and the color of the rock, use the Igneous Rocks Identification Table to name the rock. Write down all the information in the Igneous Rocks Worksheet.

To use the Igneous Rocks Identification Table, simply intersect the column that represents the color (type of igneous rock that reflects mineral composition) with the horizontal row that represents the texture and the intersection point will give you the name of the igneous rock.

Example

Assume that a rock contains pink-colored minerals among other minerals with different colors so using guideline (2A-2) then it is possible that the rock is acidic/felsic. Furthermore, if you see minerals covering the whole sample, then it is coarse grained rock, that is, it has phaneritic texture according to the guideline (1A).

Intersecting the vertical column (acidic/felsic) with the horizontal row (phaneritic) in the Igneous Rocks Identification Table will give you the name of the rock and in this case it is granite.

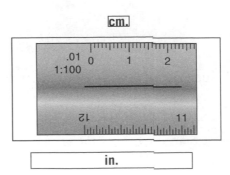

Figure 4.1

Index (A): General colors for each group:

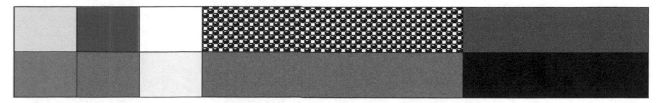

Index (B): Detailed colors for each group:

Pink to low pink	Low pink to light purple	No pink No purple
Light gray	Light gray to Intermediate gray	Dark gray
White		Black
Yellow to light yellow	Very pale light yellow	No yellow
Red to dark red	No red	Brown
No green	Pale green to pale brown	Dark green
Tan to salmon		No tan or salmon

Index (C): Three major colors, shades and percentage of dark minerals in each group:

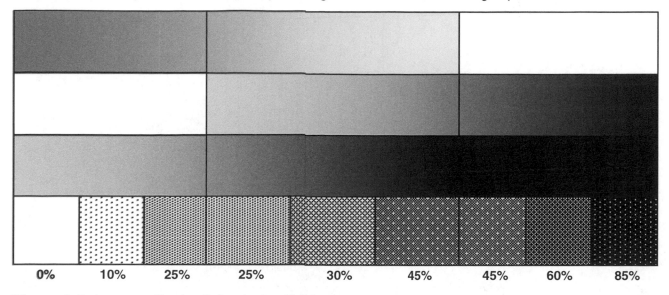

| 0% | 10% | 25% | 25% | 30% | 45% | 45% | 60% | 85% |

Figure 4.2 Igneous Rocks Color Index Tables

IGNEOUS ROCKS

Igneous rocks are classified according to:
(1) Mineral composition.
(2) Texture (grain size).

Modified by Nabil Kanja

*Pegmatite represents an end product of magma crystallization, rich in rare elements volatiles.

		Acidic or Felsic — Light Color Rocks — Silica Oversaturated	Intermediate	Basic or Mafic — Dark Color Rocks — Silica Undersaturated
Minerals	**Mineral Composition**	Potassium Feldspar (Microcline/Orthoclase) — Quartz	Plagioclase Feldspar — More Na Plagioclase (e.g. Albite) — Biotite — Amphibole	More Ca Plagioclase (e.g. Labradorite) — Pyroxene — Olivine
				100% / 75% / 50% / 25% / 0%

Texture

Rocks	Texture		Felsic	Intermediate	Mafic
Extrusive or Volcanic	Pyroclastic	Tuff (fragments less than or equal to 2 mm) / Volcanic Breccia (fragments greater than 2 mm)			
	Glassy	Obsidian (massive) → Do the test			
	Cellular	Pumice (frothy and vesicular)		Scoria (vesicular)	
	Aphanitic (less than 2 mm)		Rhyolite	Andesite	Basalt
I + E	Porphyritic	C and F grains	Rhyolite or Granite	Andesite or Diorite	Basalt or Gabbro
Intrusive or Plutonic	Phaneritic (between 2 mm and 10 mm)		Granite	Diorite	Gabbro
	Pegmatitic (more than 10 mm)		Granite Pegmatite	Diorite Pegmatite	Gabbro Pegmatite

Fine-grained / Coarse-grained

1: Pegmatite Granite

2: Porphyritic Granite

3: Granite

4: Pink Granite

5: Rhyolite

6: Rhyolite

© Tyler Boyes/Shutterstock

Figure 4.3 (*Continued*)

7: Pumice

8: Pumice

9: Diorite

10: Diorite

11: Andesite

12: Andesite

Figure 4.3 *(Continued)*

13: Andesite

14: Gabbro

15: Basalt

16: Scoria (brown)

17: Scoria (red)

18: Scoria (black)

Figure 4.3 (*Continued*)

19: Obsidian

20: Obsidian

21: Peridotite (green)

22: Peridotite (black)

23: Peridotite (gray-green)

24: Volcanic Tuff

Figure 4.3

Foliation Texture (Types) for Foliated Metamorphic Rocks

Foliation means planar or linear orientation and this could be planar minerals (mica, chlorite, kyanite, serpentine, and talc), linear minerals (hornblende and staurolite) or planar/linear structures. Foliation is divided to different types illustrated as figures seen below. Foliation texture is produced by differential stress (accompanied with heat) and the alignment of platy or prismatic crystals to direction perpendicular to applied stress.

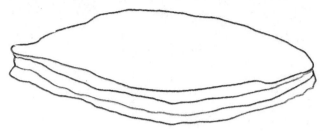

Slaty foliation (texture): shows layers and the minerals cannot be seen with naked eye. This type is seen in slate and phyllite due to recrystallization of platy minerals like mica and chlorite to direction perpendicular to stress (pressure). Note that slaty foliation cuts through bedding planes.

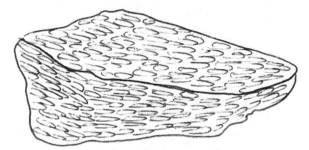

Schistose foliation (texture): shows larger crystals (minerals) that could be seen by naked eye and there could be layering or not. Shiny luster usually characterizes schist foliation. Mica schist, garnet schist, and chlorite schist show this type of foliation. One type of schist foliation is shown in this figure by prismatic crystals of the mineral amphibole (hornblende) going one direction. Example of this type of foliation is amphibolite (amphibole schist).

Phyllitic foliation (texture): shows layers with wavy (crenulated) surfaces. This type of foliation has more shiny surfaces (sheen) than slaty foliation. Minerals are larger than slaty foliation but cannot be seen with naked eye. Phyllite shows this type of foliation. Mica, chlorite, and graphite give sheen and cleavage.

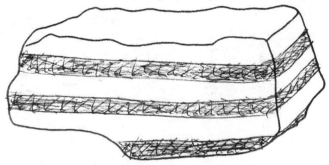

Gneissic foliation (texture): shows alternating bands of black and white or light and dark minerals. Formed from separation of dark-colored minerals from light-colored minerals due to high grade of metamorphism. Gneiss shows this type of foliation.

Figure 5.1

Table 5.2

	Nonfoliated to Weakly Foliated Metamorphic Rocks	
Name	**Chief Minerals**	**Other Properties and Notes**
Quartzite	Quartz	Quartzite could be white- or pink-colored rock. Quartzite is harder than marble and doesn't react with HCl acid as marble does. Quartzite is nonfoliated, however some times, is foliated with white and pink or purple banding. It is possible to see the crystals as grains, which could be coarse to medium grains. Produced by contact metamorphism (also could be produced by other types of metamorphism). Parent rock is sandstone or arkose.
Marble	Calcite, dolomite	Color could be white, pink to purple. Nonfoliated, although sometimes you could see like banding of brown to green color, which could be biotite. Some times limestone alternating with shale, when metamorphosed, could give banded marble. You can see the coarse grains. Produced by contact metamorphism (also could be produced by other types of metamorphism). Parent rock is limestone or dolomite.
Anthracite	Carbon	Black color and shiny. Some times shows layering. Anthracite looks like bituminous coal; however, anthracite has yellow shiny surface, whereas bituminous coal has black shiny surface. Produced by low- to medium-grade metamorphism of coal (bituminous, lignite).
Meta-conglomerate	Rock fragments and quartz	Produced by metamorphism of conglomerate. Looks like conglomerate but some times with starched or flattened pebbles (weak foliation). Also if metaconglomerate is broken, fractures go through pebbles.
	Foliated Metamorphic Rocks	
Slate	Mica, chlorite with quartz	Slate could be red, black or gray, and green. The black slate could be easily confused with black shale, because both look same in color and layering; however, slate is less blackish, more shiny and if you scratch it with steel file it will give you white scratch, whereas black shale will give you black scratch. Produced from regional metamorphism of shale or mudstone. Considered low grade of metamorphism.
Phyllite	Mica, chlorite	Some times black but more shiny than slate and some times greenish black with wavy surface (crenulations.) Produced by regional metamorphism of shale. Also produced by metamorphism of slate. Considered low to intermediate grade of metamorphism.
Chlorite schist	Main mineral: chlorite Other minerals: chlorite, mica, quartz, amphibole	Like mica schist shiny but green colored. Produced by regional metamorphism of basalt. Also produced by metamorphism of phyllite. Considered intermediate to high grade of metamorphism.
Mica schist (sometimes garnet schist)	Main mineral: mica Other minerals: quartz, Amphibole and sometimes Garnet	Like chlorite schist but white, brown to black colored sheets. Sometimes it contains the mineral garnet, shown as large red crystals. Produced by regional metamorphism of shale or rhyolite. Also produced by metamorphism of phyllite. Considered intermediate to high grade of metamorphism.
Amphibole schist (amphibolite)	Main mineral: Amphibole Other minerals: Quartz, Mica	Black colored and easily recognized by its shiny needle shaped crystals of amphibole (hornblende.) Produced by regional metamorphism of basalt, gabbro or diabase. Also produced by metamorphism of phyllite. Considered intermediate to high grade of metamorphism. Some times it is possible to see veins of quartz in it.
Gneiss	Quartz & Feldspars (light minerals) with Biotite and Hornblende (dark minerals)	Is shown as black and white alternating bands or colored bands, although some gneisses show blacker banding than white and in this case, we call it amphibole (hornblende) schist. Produced by regional metamorphism of shale, granite or diorite. It is considered high grade of metamorphism.

The following notes and observations are important in the identification and understanding of metamorphic rocks:

TEXTURE

|||

1. Foliation: Rocks showing parallel or linear arrangement of minerals and these mineral are interlocking (nearly equal size grains and no spacing between grains).

2. Some nonfoliated rocks may show type of foliation. Figure (5.2-1, 3) shows marble with a faint foliation. This is due to the existence of impurities like biotite and actinolite that show a weak orientation, layering or banding inherited from parent rock, limestone. Marble also could show streak(s) of organic material that will look like foliation. Quartzite, like marble, may show weak foliation due to the existence of inherited impurities like muscovite and kyanite or the existence of structures like cross bedding or ripple marks from metamorphosed sandstone.

   Metaconglomerate is another rock that shows weak foliation as pebbles are flattened or elongated (stretched) due to deformation.

   It is important to say that some metamorphic rocks like marble, quartzite, and metaconglomerate may show some foliation however these examples are considered nonfoliated rocks. Foliation is texture that develops due to differential pressure associated with mountain building processes and not due to temperature or to a less degree confining (lithostatic) pressure. Furthermore, foliation requires platy (flat), elongated minerals as the texture refers to the parallel/linear alignment (arrangement) of these minerals. Any metamorphic rock that shows type of foliation and will not satisfy the above two requirements then it is not considered foliation.

   Note that in Table (5.2), the rocks are divided to two groups; nonfoliated to weakly foliated metamorphic rocks and foliated metamorphic rocks and as mentioned before weakly foliated metamorphic rocks are not considered foliated metamorphic rocks.

3. Texture terms like slaty, phyllitic, schistose, and gneissic are used; however, it is possible to confuse textures like schistose and gneissic because they resemble each other in some of the foliated metamorphic rocks. Look at Figure (5.2-16, 20). To differentiate between the two textures, gneiss is more compact and will not break along foliation whereas schist mostly will break along foliation and is sometimes flaky. The reason for this similarity of foliation between schist and gneiss is related to the platy minerals and the existence of large crystals with the presence of minerals like quartz and feldspar (not platy) in schist. Also the existence of platy minerals with non-platy minerals causes schist to split along irregular surfaces that are parallel to foliation.

4. Slaty texture is also referring to the rock that splits (cleave) along parallel planes and therefore giving slaty cleavage and the cleavage here is referring to the alignment of platy minerals that give planes of weakness and these planes of weakness are not the bedding planes in sedimentary rocks or planes of weakness in minerals (not related to internal atomic structure), whereas phyllitic texture is referring to the rock that will split along planes that are wrinkled, and therefore, the cleavage for phyllite is not as clear as slate.

5. Slate may look like shale in that both are black and show layering however slate splits across bedding planes whereas shale splits along bedding planes.

6. Granoblastic: Rocks with random interlocking grains. Marble and quartzite are two examples.

7. Porphyroblastic: This type of texture could be found in foliated and nonfoliated metamorphic rocks. This type of texture refers to large crystals surrounded by small crystals (like porphyritic texture in igneous rocks) and the porphyroblasts are elongated (eye-shaped) crystals.

   This type of texture develops when growth of some minerals are faster than other minerals. For example garnet-mica schist is schist with garnet porphyroblasts and mica (both muscovite and biotite) surround the garnet.

8. Minerals that show foliation include minerals that could be platy minerals like mica, chlorite, talc, serpentine, and kyanite; prismatic minerals like amphibole, serpentine, and sillimanite; fibrous minerals like serpentine.

MINERAL COMPOSITION AND NAMING OF METAMORPHIC ROCKS

1. Index minerals are minerals found in metamorphic rocks that indicate grade of metamorphism hence used to determine temperature and pressure of metamorphism and include chlorite, biotite, garnet, staurolite, kyanite, sillimanite. Look at Figure (5.3).

   Figure (5.3) shows index minerals (different colored bands) at range of temperatures. Note that the light blue bands of quartz, feldspars, and muscovite represent minerals that are stable under wide range of temperature and pressure and, therefore, they are not index minerals. Furthermore, the temperatures for metamorphism starts at approximately 200°C and ends at approximately 800°C and any temperature less than 200°C are covering area before metamorphism (sedimentary rocks) and temperatures more than 800°C are covering area of melting (magma formation) and the formation of igneous rocks.

2. In general, the index minerals are arranged of increasing temperature starting with lowest temperature, chlorite moving to muscovite then garnet then staurolite then kyanite then sillimanite as the end member representing high grade of metamorphis (highest temperature). Some minerals like blue amphibole, green pyroxene (omphacite), and garnet (rich in iron, magnesium) are high pressure minerals.

3. Furthermore, blue amphibole is an indication of high pressure but low temperature and green amphibole is an indication of high pressure and high temperature.

4. Clay forms at surface and stays stable until temperature is in the range of 200–250°C. Clay changes to mica and chlorite.

5. Minerals like quartz, feldspars, and calcite can't be used as index minerals since they cover wide range of temperature that is they are stable over wide range of temperature and pressure.

6. The name of mineral(s) added to the name of the rock could be the name of the dominant or distinctive (not abundant) mineral. For example, mica schist and garnet mica schist. Prefixes should be placed in order of increasing abundance. For example biotite-quartz-plagioclase gneiss contains more plagioclase than quartz and more quartz than biotite.

7. The prefix "meta" sometimes are added to the parent rock to give the name of the metamorphic rock. For example, metaconglomerate, where the sedimentary rock conglomerate metamorphosed (changed) and the change was shown as stretched, deformed, and fused pebbles. Look at Figure (5.2-7).

1: Pink marble with weak foilation

2: Light pink marble

3: White marble with gray coloration

4: White marble

5: Quartzite

6: Quartzite

Figure 5.2 (*Continued*)

7: Metaconglomerate

8: Anthracite

9: Red slate

10: Green slate

11: Black slate

12: Black phyllite with wavy surface

Figure 5.2 (*Continued*)

13: Black phyllite with wavy surface

14: Purple phyllite

15: Light green phyllite

16: Mica schist

17: Mica schist with garnet

18: Mica schist

Figure 5.2 (*Continued*)

19: Black and white gneiss   20: Colored (pink) gneiss

Figure 5.2

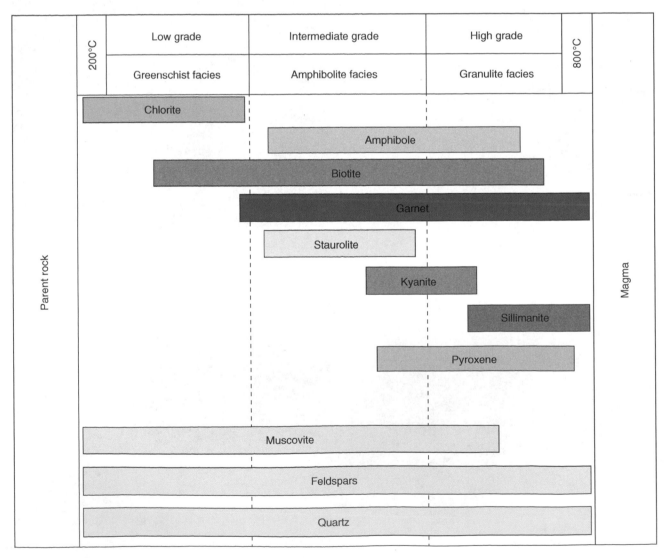

Figure 5.3 Index Minerals and Grade of Metamorphism

MASS WASTING

If an object was put on an inclined surface, that object will be under the influence of two forces; one force is shear stress (Sstrs) and that force will pull any object down slope under the influence of gravity. The second force is shear strength (Sstn), and that force opposes and works against the shear stress due to friction. Look at Figure (6.1-1).

If shear stress is less than shear strength then the object will stay stable and there will be no motion. Look at Figure (6.1-2). However, if shear stress is more than shear strength then the object will not be stable and it will move down slope. Look at Figure (6.1-3).

Therefore, mass wasting refers to the down slope movement of materials under the influence of gravity.

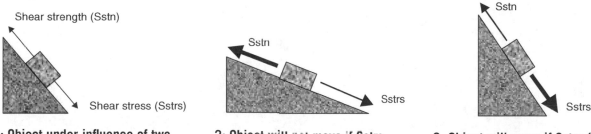

1: Object under influence of two forces

2: Object will not move if Sstrs (thin arrow) is less than Sstn (thick arrow)

3: Object will move if Sstrs (thick arrow) is more than Sstn (thin arrow)

Figure 6.1

FACTORS INFLUENCING MASS WASTING

Mass wasting could be triggered by different factors and agents that could include the following:

1. Angle of repose
2. Vegetation
3. Water
4. Shape and size of grains
5. Rocks type and structure
6. Earthquakes and volcanic activity
7. Weight addition

1. Angle of repose is the maximum angle at which the slope will be stable, that means objects (like sediments) on the slope will not move. For example, if angle of repose for specific slope is 45° then increasing the angle of repose beyond 45° will make the slope unstable (objects, like sediments, on slope will move down); however, decreasing angle of repose to less than 45° will still make the slope stable (objects, like sediments, on slope will not move down). Different materials have different angle of repose and angle of repose could range from 0° to 90°. Table (6.1) shows angle of repose for different materials. Please note that these values may vary from the values in the table.

Table 6.1

Material	Angle of Repose
Ash	40°
Aspalt (crushed)	30°–45°
Chalk	45°
Clay (dry)	25°–40°
Clay (wet)	15°
Coal (industrial)	35°–38°
Earth (soil): dry	20°–45°
Earth (soil): moist	25°–45°
Earth (soil): wet	25°–30°
Earth (soil): in general	30°–45°
Granite	35°–40°
Gravel (dry)	30°–45°
Gravel (round to angular)	30°–50°
Gravel with sand	25°–30°
Salt	35°
Sand (dry)	34° also (20°–30°)
Sand (moist)	30°–45°
Sand (wet)	45° also (20°–45°)
Sand and clay	20°–35°
Snow	38°
Sand (water filled)	15°–30°
Sugar	35°–37°

2. Vegetation like plants could make the slope more stable because the roots act like a network that traps sediments and will not allow the soil to drift.

Plants also could act like a barrier that blocks the water from running off the surface and therefore infiltration will be more than run off and this reduces soil erosion. Although this may help the soil to be stable but at the same time infiltration of water in soil could cause that soil to be saturated. Making the soil saturated not only adds weight to it but also lubricates the soil leading to mass wasting.

3. Water is critical agent but it is not necessary for mass wasting to happen.

For example adding little amount of water to soil could make the soil cohesive and stable but also adding large amount of water could make the soil change from solid state to liquid state because water will break the bonds between soil particles and cause mass wasting. Furthermore, water acts as a lubricant agent that eases the initiation of mass wasting.

Shape of grains and size	Stability
 Angular fragments	Angular grains are more stable than rounded grains.
 Rounded fine fragments	Fine grains are more stable than coarse grains (assuming both have the same shape).
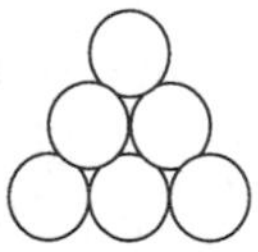 Rounded coarse fragments	Coarse grains are less stable than fine grains of the same shape.
 Mixed angular fragments	Grains of different sizes that have angular shape are more stable than grains that have same size and have rounded shape.

Figure 6.2

4. It is known that angular grains are more stable than rounded grains (assuming both have nearly the same size). Also fine grains are more stable than coarse grains (assuming both have the same shape). Look at Figure (6.2).

5. Generally speaking sediments (loose rock fragments) are less stable than solid rocks and solid rocks on the other hand must be investigated concerning their properties, characteristics and type. For example shale, limestone, dolomite (dolostone), slate, and marble are some of the rocks that due to their properties are less stable than other rocks.

Shale cracks or fractures (contracts) when it is dry and swells (expands) when water is added to it. Furthermore, shale is made of layers (planes of weakness).

Limestone and dolomite are rocks that dissolve when hydrochloric acid is applied (for example, acid rain, one type of chemical weathering) and although marble is a solid compact dense rock but it will dissolve like limestone, because both are made of the mineral calcite.

Slate is made of slabs that can be easily broken under pressure.

Features like bedding planes, fractures and faults, foliation and the direction of such features could weaken the rock. The layers (bedding planes) in Figure (6.3-1) are pointing to the same direction of the slope so mass wasting is more possible than the situation in Figure (6.3-2) where the layers (bedding planes) are in a direction that is opposite to the direction of the slope.

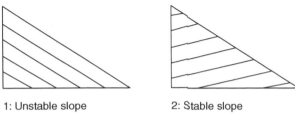

1: Unstable slope   2: Stable slope

Figure 6.3

6. Earthquakes and volcanic activity increase fractures in specific area and that cause an increase in weakness of rocks. Furthermore seismic and volcanic activity causes vibrations and if there are loose sediments (not solid rocks) then the loose sediments will act as amplifiers to these vibrations and vibrations will increase in intensity.

Seismic activity could cause liquefaction and this happens to soil that have a lot of water (wet soil) or saturated soil and because of seismic vibrations the soil behaves like a liquid. It is important to mention that liquefaction happens by severe shaking(s) or repeated shakings and soils with nearly the same grain sizes are more susceptible to liquefaction than a soil with different grain sizes (mixed). Also rounded grains are more susceptible to liquefaction than angular grains. Furthermore, fine grains are less susceptible to liquefaction than coarse grains.

Clayey soils could show similar behavior to liquefied soil but do not liquefy the same way as sandy soils do.

7. Overloading by adding weight to any slope increases the instability and increases the risk of mass wasting. For example, adding a swimming pole not only may cause water to leak in to the bedding planes but also puts pressure on soil or rocks and widens the possibility of fractures development by decreasing the strength of the rock.

MASS WASTING CLASSIFICATION

Mass wasting is classified to different types based on type of materials, rate of motion, and type of motion. Sometimes different types of mass wasting could merge together or produce other types of mass wasting.

Table (6.2) explains in a very simple way the criteria used for classifying mass wasting. Read it carefully since it will help you to understand the different types of mass wasting.

Table 6.2

Type of Material	Type of Motion	Rate of Motion
Wet or dry	Existing surface or free fall	Fast, slow or very slow
Loose sediments or solid rock (consolidated)	Type of surface (flat or curved)	
Cohesive (silt, clay) or noncohesive (sand)	Motion as one unit or discrete parts	

Rock characteristics:
  – Type of rock (shale, slate, limestone, etc)
  – Soluble or nonsoluble
  – Fractures, faults, foliation, and bedding
   planes and their orientation
	Motion along planes of weakness like bedding planes or fault planes	

Shape and size of rock fragment (grains):
  – Rounded or angular grains
  – Fine, coarse or mixed grains		

TYPES OF MASS WASTING

There are different types of mass wasting and the following are some of the types and their characteristics. It is important to say that one type of mass wasting could trigger or even change to another type of mass wasting and sometimes the mass wasting is combination of more than one type (complex).

Although table (6.2) is used as a simple way to classify mass wasting it is important to know that the term "flow" is used when water is involved in mass wasting and the term "slide" is used when water is not involved. In addition, looking at the different velocities of mass wasting it is helpful to refer to rate of motion in general. For example, avalanches are mass wasting with velocity more than 5 km/hr, whereas rock slide, mud flow, and slump are mass wasting with velocity more than 1 km/hr. Earth flow and debris flow are mass wasting with velocity more than 1–10 m/hr, and creep and solifluction are slow type of mass wasting with velocity more than 1 cm/yr.

Avalanche

– The motion is very rapid and the material could be mixture of rock fragments of different sizes, air, and sometimes water where the air (that is trapped under rock mass) acts as a cushion and this could lead to high velocities (201.17 kilometer per hour or 125 miles per hour) and wide area coverage therefore it is the fastest and most destructive type of mass wasting.

– No motion of soil is involved.

– Is triggered by earthquakes, volcanic activity, and weathering and weakening of rocks.

Rock Fall

– No surface

– Free fall

– Happens when frost wedging for example happens at top of a steep cliff and fractures develop and rock fails suddenly. Rock fragments that accumulates at base of cliff are known as Talus, and these rock fragments are angular, coarse, and porous with angle of repose = 45°. Look at Figure (6.4).

– The material is usually dry and not cohesive.

– No motion of soil is involved.

– The rate of motion is fast.

– This type of mass wasting is seen a lot at highway and road cuts.

Rock Fall

Figure 6.4

Source: USGS.

Rock Slide

– There is surface

– The motion of material is as one unit like the motion of bedrock along planes of weakness (like bedding planes, fractures, faults, or foliation). The motion of bedrock as one unit could evolve to smaller units when the major unit mass starts breaking with advancing movement. Look at Figure (6.5)

– Rate of motion is very fast

– Very destructive

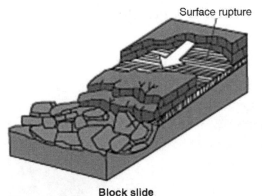

Surface rupture

Block slide

Figure 6.5

Source: USGS.

Mudflow

– There is surface

– Material includes sediments, usually fine (silt and clay) to medium sized fragments (sand) and water that is less than 60% therefore material is less cohesive (because of sand). Look at Figure (6.6).

– Usually motion happens to soil or regolith

– Happens in arid to semi-arid climate in mountainous regions with sparse vegetation where sudden and intense rain fall saturates loose sediments and triggers this type of motion.

– Rate of motion is up to 129 kilometers per hour (80 miles per hour).

– Lahar is considered one type of mudflow and is combination of volcanic ash and water.

Mudflow

Figure 6.6

Debris Flow

– Material includes sediments of coarse sized fragments (usually if more than half of the fragments are coarser than sand) and water that is less than the amount of water in mudflow. Material is non-cohesive and could include trees, cars, and structures.

Earthflow

– There is surface

– Material includes less water than mudflow.

– Material is more cohesive and includes more silt and clay and less sand

– The end of earthflow is usually rounded (shape)

– Motion of material is defined by lateral boundaries and is parallel to surface and not rotational as slump. Look at Figure (6.7)

– Rate of motion is from (1) millimeter per day to meters per day.

– Happens in humid climate

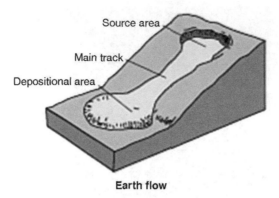

Earth flow

Figure 6.7

Source: USGS.

Slide and Slump

– These two terms are used with different types of mass wasting and they indicate the shape of the surface.

– Slide has nearly flat surface and slump has more curved surface.

– Slump indicates motion of material along a spoon shape or sloping bowl surface and the motion of material, like land slide, includes the surface and the ground (material below surface) however slump is usually rich in clay. Slump could be produced when the base of the cliff is removed by wave action or stream action or road and building construction and the shape of the cliff or scarp produced is crescent shaped. Look at Figure (6.8).

– Material in slide and slump is drier than the flow type of mass wasting.

Figure 6.8 Slump

Source: USGS.

Creep

– There is surface

– Happens due to daily freezing and thawing or dry and wet conditions

– The material that moves is soil saturated with water that expands vertically when it freezes, and shrinks diagonally when it thaws, so soil expands up vertically and shrinks diagonally. This could happen in cold climate and also could happen in environment that experiences wet and dry cycles

– The motion happens in upper few feet of soil

– Some indications of creep are tilted trees, fens, electric posts and the development of fractures in basements. Look at Figure (6.9)

– Rate of motion is very slow (slower than solifluction)

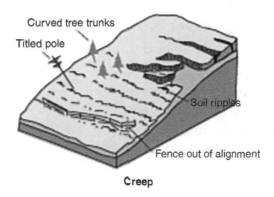

Figure 6.9

Source: USGS.

Solifluction

– There is surface

– Happens in polar regions where there is permafrost and the soil is frozen the whole year. The active layer (upper layer of permafrost) melts during summer and moves down slowly and freezes during winter (no motion). Look at Figure (6.10)

– Rate of motion is very slow (faster than creep)

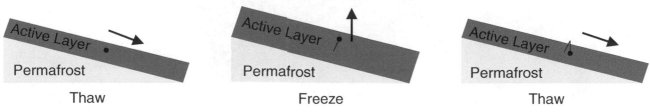

Figure 6.10

PART (A): FINDING ANGLE OF REPOSE

Bring three types of materials and assign them letters A, B, and C and write their characteristics in the following table, Table (6.3):

Table 6.3

Type of Material (Name)	Letters	Characteristics			
		Color	Grain Size	Grain Shape	Sorting
	A				
	B				
	C				

Determine the angle of repose for the three types of materials individually by putting each material on inclined surface, like clipboard, with protractor (to measure angle) attached to the clipboard. Look at Figure (6.11). In order to measure the angle of repose for each material, start by putting the specific material on the surface (clipboard) in horizontal position. Put the specific material on the horizontal surface (clipboard) then start tilting the surface (clipboard) slowly and gradually and as soon as the specific material starts moving (sliding), stop and take the value of the angle before sliding as the angle of repose. You have to be patient and may need to repeat it many times to get the angle. Notice the angle of repose in Figure (6.11) is approximately 25°.

Put the obtained values for angle of repose for each material in Table (6.4).

Figure 6.11

Table 6.4

Type of Material	Letters	Angle of Repose
	A	
	B	
	C	

1.A) Out of the three materials, which one, has the highest angle of repose?

2.A) Explain your answer for question (1.A) that is say why?

3.A) Consider these two situations. What would be the influence of water if it was added to any of the three materials? Also what would be the influence of water if it was added to the surface (clipboard)?

PART (B): MEASURING ANGLE OF REPOSE

The angle of repose could be obtained from triangle if vertical drop and horizontal distance is known. Look at Figure (6.12).

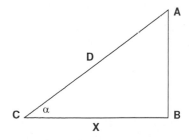

Figure 6.12

The slope (D) is equal to vertical drop (A-B) divided by horizontal distance (X). On topographic maps, vertical drop is the elevation at (A) subtracted from elevation at (B) and the horizontal distance is the distance that you measure on map using the specific scale of the map and notice that this distance, which is (X) is not the distance (D) between (A) and (C) since this distance (D) is distance of slope and not horizontal distance between two points.

Let us assume that elevation at point (A) is 100 ft and elevation at point (B) is 50 ft and the horizontal distance (X) is 130 ft. Then calculating slope (D) and angle of tilt is as follows:

Slope = (A − B)/X

= (100 ft − 50 ft)/130 ft = 0.3846 and if this number is multiplied by 100 then it is = 38.46%

Now it is possible to obtain the angle of tilt (α) from:

Tan (α) = (A − B)/X = 0.3846, then the angle (α) = $\tan^{-1}$ (0.3846) = 21.03°.

Also it is possible to know the actual distance (D) from Figure (6.12) or the hypotenuse distance from:

$(D)^2 = (X)^2 + (A - B)^2$

$(D)^2 = (130)^2 + (50)^2 = 14400$

$(D) = \sqrt{14400} = 120$ ft

From the previous example, the slope can be expressed in different ways. Slope is steepness of slope so slope can be gentle slope or steep slope. Also slope is angle of tilt or inclination so looking at Figure (6.12) slope could be expressed as angle in degrees. Furthermore slope is the ratio between vertical drop and horizontal drop so can be expressed as number (decimal) or as percentage.

Using Figure (6.12) and assuming that the vertical drop (A-B) is equal to 45 ft and the horizontal distance (X) is 30 ft, answer the following questions:

1.B)   Express the slope as ratio and do not reduce it to any lower form.

2.B)   Express the slope as a decimal.

3.B)   Express the slope as a percentage.

4.B)   Find the angle of slope (tilt) using the answer obtained earlier from question (2.B). Show your calculations.

5.B)   Look at Figure (6.13) and notice the sign that is showing a slope with a number above the slope and the number is 10% and this is same as saying on some signs 10% downgrade. Furthermore notice that below the triangular sign there is another sign (rectangular) that says "Low gear." Also if you look at the picture carefully you will notice that the area is mountainous.

What is a 10% slope when expressed in degrees? Show your calculations.

Figure 6.13

PART (C): ANGLE OF REPOSE IN THE FIELD

Go outside and try to measure the angle of repose for one of the cliffs (or if hill then one side of the hill). Use the same method in part (B) by using simple trigonometry. Figure (6.14) is showing a cliff and above it a triangle. This triangle is to help you calculate the angle of slope and notice that the triangle (gray) is upside down.

Climb the top of the cliff and measure the elevation at point (A) and do same thing and measure the elevation at point (B). Then measure the distance between point (A) and point (B) represented by the irregular line of the cliff and this distance represents the slope distance between point (A) and point (B). Knowing the vertical drop (elevation at point (A) – elevation at point (B)) and knowing the slope distance between point (A) and point (B) then it is possible to obtain the angle of slope, which is angle (α).

Notice that it is possible to measure the elevation either by using a GPS system or using topographic map. Furthermore it is possible to measure the distance between any two points by using a map or using a measure tape, however you should be careful whether this distance is the horizontal distance between two points or the distance of slope between two points. Concerning this part, you need to now the slope distance or the distance represented by hypotenuse of triangle or cliff as figure (6.14) illustrates. Also notice that angle (α) for cliff is the same as angle (β) for triangle and that the angle (α) for cliff is obtained from Sin (α) = Opposite/Hypotenuse.

Elevation at point (A):

Elevation at point (B):

Elevation at point (A) Elevation at point (B):

Slope distance between point (A) and point (B):

Angle of slope (α):

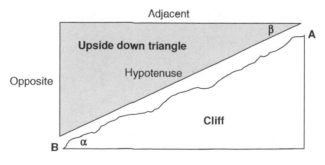

Figure 6.14

PART (D): PROBLEMS RELATED TO MASS WASTING

The following questions could have more than one answer and the answers should be related to figures. Do not assume facts that are not there or facts that are not seen in these figures, unless it is a possibility:

1.D)   Look at Figure (6.15):

   a.   Determine the type of this mass wasting.

Fill th

b.   How can you make this type of mass wasting stable?

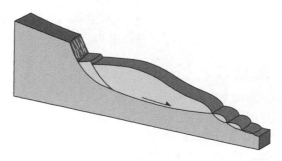

Figure 6.15

2.D)   Look at Figure (6.16):

a.   There is a horizon (rock layer or bed) saturated with water and this could cause mass wasting. What is this type of mass wasting?
Notice that you should consider the layer that is saturated with water and the existence of the bedding plane (plane of weakness) between the layers.

b.   How can you make this type of mass wasting stable?

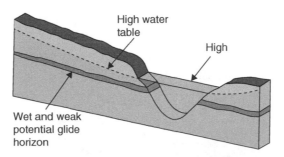

Figure 6.16

3.D)   Look at Figure (6.17):

a.   Determine the type of this mass wasting at location (A).

b.   How can you make this type of mass wasting stable?

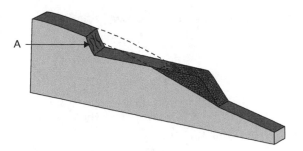

Figure 6.17

4.D) Look at Figure (6.18-A) and (6.18-B):

Figure (6.18-A) shows unstable slope that have been changed to stable slope as shown in Figure (6.18-B).

Do you think that this is a good solution? Say yes or no and mention why?

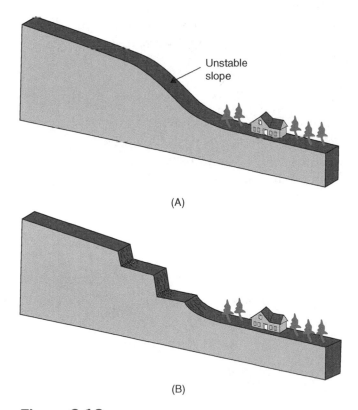

(A)

(B)

Figure 6.18

5.D)   Look at Figure (6.19) carefully and then answer the following questions:

   a.   Which one is more stable Figure (A) or Figure (B) and why?

   b.   Is the slope in Figure (C) stable or not? Why?

   c.   How can you make the slope of Figure (D) stable?

   d.   In Figure (E) there is a truck unloading boulders and gravel. Is this increasing or decreasing the stability of the slope? Why?

   e.   In Figure (F) there is a rockslide or rock fall. At the base (foot) of the cliff (mountain) there is a road. Suggest or list three solutions that can be used or applied to protect that road from any future mass wasting. Be creative.

   f.   Look back again at Figure (B) and notice that the tilting of the layers are in direction different than the direction of the slope (cliff). This makes the slope stable. However it is possible that even in these conditions the slope may experience mass wasting. Mention the name of the possible mass wasting and explain how that specific type of mass wasting could happen.

   g.   Look at Figure (C), then answer the following question with either true or false:
        Since the tilting of the layers is in the same direction of the slope, therefore the slope is stable. (True, False)

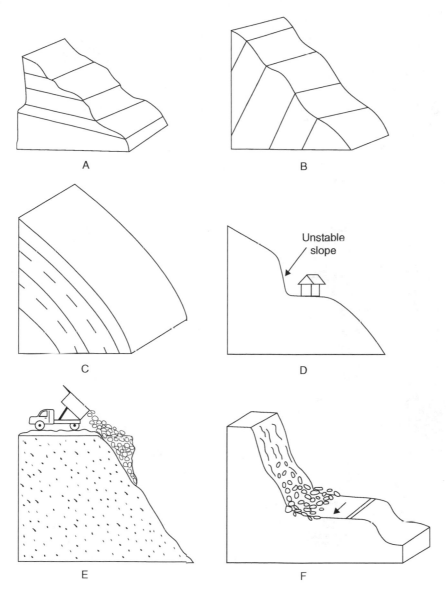

Figure 6.19

6.D) Imagine that you are an engineer or consultant for a construction company and you are trying to build or design a stable slope at the sides of a highway. Assume that you are given the following sheet, shown as table (6.5), that includes information about different materials, suggestions and solutions. Try to use your knowledge and judgment to select the proper materials, suggestions, conditions and solutions to build a safe stable slope. For each row (not column) on table (6.5) circle your option(s) and selections. You could circle more than one option for each row.

Table 6.5

Shale	Conglomerate	Slate	Limestone	Breccia	Granite
*Shale has angle of repose = 20° *Expands and contracts with addition and removal of water *Shows layering	*Conglomerate has angle of repose = 40° *Made of rounded fragments	*Slate has angle of repose =25° *Shows foliation	*Limestone has angle of repose = 45° *Dissolves with acidic rain *Sometimes fractured	*Breccia has angle of repose = 40° *Made of angular fragments	*Granite has angle of repose = 40° *Made of stable minerals at surface conditions *Sometimes fractured
Plant trees	Plant grass	Plant flowers	Plant bushes	Don't plant anything	
Select rounded grains of the same size	Select angular grains of the same size	Select rounded grains of different sizes	Select angular grains of different sizes	Add weight to soil (like structures or swimming pool)	Reduce weight that is exerted on soil
Put retaining walls at the base of slope	Put net to cover the slope	Build tunnel to cover that part close to slope	Select area that will not vibrate from earthquakes and volcanoes	Select area close to streams and/or close to earthquakes and volcanoes	Divide a steep slope to terraces (short steps)
Add more water to soil to reach saturation	Reduce amount of water in soil to just keep it moist	Keep the soil dry by draining it using pipes			
Increase angle of slope beyond angle of repose	Decrease angle of slope below angle of repose	Do not worry about angle of repose for slope			

TOPOGRAPHIC MAPS

Any map in order to be considered a real map, it should include three things:

1. Legend (key)
2. Scale
3. North direction

LEGEND/KEY

Shows the meaning of the colors and the symbols used on the topographic map, for example:

Black color = structures (buildings, bridges, parking lots) and roads

Green color = vegetation

Red color = high ways and main roads

Blue color = water (lakes, ponds, streams)

Brown color = elevation (contour lines)

Purple color = expansion (new additions)

Pink color = urban areas

Some of the symbols used in topographic maps are shown in Figure (7.16)

SCALE

Is the ratio of the size on the map to the size on ground and there are three types of scales:

1. Graphic/bar scale
2. Ratio/representative fraction (R.F) scale
3. Verbal scale

It is important to say that the larger the scale, the smaller the area covered and the more detailed the map is. For example, if we have two maps, map (A) has scale 1:1000 and map (B) has scale 1:10,000 then map (A) has larger scale since 1/1000 = 0.001 whereas map (B) will have smaller scale since

1/10,000 = 0.0001. Therefore, map (A) will cover smaller area but will show more details of the area covered, whereas map (B) will cover larger area and will show less details of the area covered.

Example

An area has been mapped at R.F. scale of 1:50,000. A new map of the same area is drawn on a scale of 1:25,000. What area of paper, in relation to the first map, will be needed for the new map?

To solve this problem first look at the first map that has the scale of 1:50,000. This map has small scale since 1/50,000 = 0.00002 compared to the second map of scale 1:25,000 since 1/25,000 = 0.00004. This means that the first map will cover larger area compared to the second map that covers smaller area.

Now using Figure (7.1) and assuming that for the first map 1 in = 50,000 in and for the second map 1 in = 25,000 in then the second map covers area 4× (times) the area covered by the first map that is the map with the scale 1 = 50,000 will contain or show, for example, four houses compared to the map with the scale 1 = 25,000. Also note that the four houses covered by the first map will show lesser details compared to the second map that will show one house but with more details.

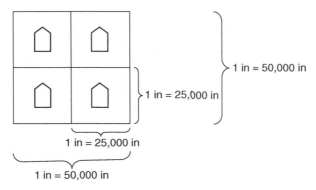

Figure 7.1

1. **Graphic/bar Scale:**

   This is shown as bar or drawn graphically so you can see it. Look at figure (7.2)

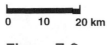

   Figure 7.2

   Graphic scale is very practical because it can be used directly to measure the distance between two points on map. Furthermore graphic scale will not shrink or expand when photocopied from the original map.

2. **Ratio/Fraction or Representative Fraction (RF) Scale:**

   When this scale is used, two rules should be followed:

   a. The scale must be written in this form: 1:20,000

      The left hand side should be always 1 so if there is a ratio scale written in this form 2:20,000 then it should be changed to 1:10,000 (the original scale was divided by 2 for both sides).

b. There should be no units written since the units on the right are same as the units on the left and this is why they are cancelled. This implies that you can use the units that you like, for example 1:30,000 means 1in on the map is equal to 30,000 in on the ground (field) or you can say 1km on map is equal 30,000 km on the ground (field) and so on.

3. **Verbal Scale:**

This scale is expressed verbally like saying 1 in = 3 miles that means 1 in on the map is equal to 3 miles in real life (ground or field).

Using this scale you should make sure that:

a. The units on the right side should always be different than the units on the left side.
b. The number on the left is always 1. For example, if the verbal scale is 3 cm = 9 km then this scale should be changed to 1 cm = 3 km (divide both sides of the original scale by 3).

Specific scales are given for specific series of USGS topographic maps. See Table (7.1):

Table 7.1

Scale	Series	1 in represents	Standard Quadrangle Size
1:24,000	7.5 min	2000 feet	7.5 in × 7.5 in
1:20,000	Puerto Rico 7.5 min	About 1667 feet	7.5 in × 7.5 in
1:62,500	15 min	Nearly 1 mile	15 in × 15 in
1:63,360	Alaska	1 mile	15 in × 20 in – 36 in
1:250,000	US	Nearly 4 miles	1° × 2°
1:1000,000	US	Nearly 16 miles	4° × 6°

NORTH DIRECTION

There are two types of north direction; one is the true or geographic north and the other one is the magnetic north. Grid north is the third type of north and is represented by a dot on top of a line. Look at Figure (7.3). Grid north will not be considered here.

1. **Geographic/true North:**

There is an imaginary line that passes through the earth (at poles) where the earth spins about it, this is called axis of rotation. The two places where the axis of rotation intersects the earth at the north and the south are referred to as the geographic or the true north and south.

Geographic/true north is shown as a star on top of a line. Look at Figure (7.3).

2. **Magnetic North:**

Is the north determined by the compass and is related to the magnetic field of the earth, therefore it changes from one location to another location and when calculations are made the magnetic north always should be corrected to the geographic or the true north and *declination* should be considered.

Magnetic north is shown as an arrow. Look at Figure (7.3).

Type of North	Symbol Used
Grid North	
Geographic/True North	
Magnetic North	

Figure 7.3

Declination

Is the angular difference between the geographic/true north and the magnetic north. Declination requires both magnitude and direction and both should be mentioned. The direction is always to the right (east) or left (west) of the geographic /true north.

The declination shown in Figure (7.4) is equal to 5° (magnitude) and the direction is to the left (west). Look also at Figure (7.5).

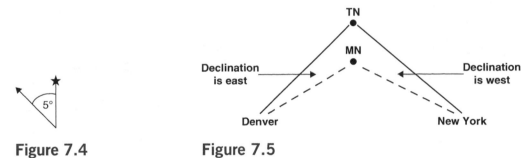

Figure 7.4   **Figure 7.5**

Topographic Map

Topographic map is a map that shows features on the surface (mountains, hills, valleys, etc.) and their elevation using contour lines.

Contour Lines

Are lines that connect points of equal elevation from mean sea level.

Contour Interval (C.I.)

Is the vertical distance between two successive contour lines.

Contour intervals could be representing large numbers (100, 250, or 500 feet) if the map covers large area or steep slopes (steep mountains) or could be representing small numbers (5, 10 or 25 feet) for small areas or gentle slope topography or if details are needed to be shown.

Contour Roles

1. Contour lines never cross or intersect. Look at Figure (7.6-1).
2. Contour lines never split. Look at Figure (7.6-2).
3. Contour lines are always continuous.
4. The fifth contour line is bold and thick and is called Index Contour Line. Usually the values are written on these Index Contour Lines. Note that, in some maps, the Index Contour Line could be the fourth contour, so pay attention to that.

5. Contour lines are shown as brown colored lines.

6. Widely spaced contour lines represent gentle slope whereas tightly spaced contour lines represent steep slope.

7. When contour lines cross a stream (valley) they bend up stream making a "V." It is possible to use this rule to determine the direction of flow for stream where the tip of the "V" points always to the opposite direction of flow for a specific stream. So for Figure (7.7) the flow of the stream is from NE to SW.

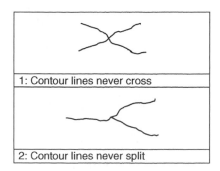

Figure 7.6

Contour Shapes

The following shapes of contour lines could represent the following features:

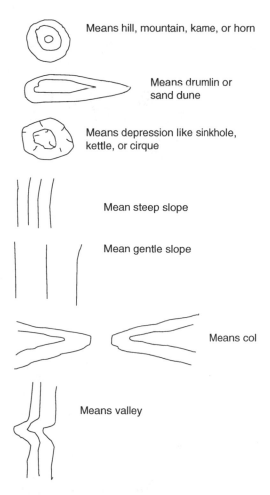

Elevation

Elevations are written with black, brown, and blue (water) color on topographic maps.

If elevation of a specific point is known accurately then these points of elevation are called bench marks (B.M.) and are shown on the map as a triangle with a central spot inside the triangle. Also letters BM and the elevation of the spot in feet are printed beside the triangle. On the ground, they are shown as numbers on rounded brass plates.

Sometimes the topographic maps show only the letters BM with the elevation without the triangle. Elevations that are less accurate are shown as a cross or X with the elevation of the spot in feet.

The elevation is obtained from the contour lines. For example and from Figure (7.7):

The elevation at point A is 700 ft

The elevation at point B is 740 ft

The elevation at point C is 770 ft

The elevation at point D is 820 ft

If the elevation point is not between two contour lines and there is no contour line passing through that specific point, like point E, then do the following:

The point E is between the contour line 820 ft and the next contour line that is not shown which is 840 ft (based on contour interval) then (840 ft − 820 ft)/2 = 20 ft/2 = 10 ft. Add the 10 ft to the lower (previous) contour line which is 820 ft so 820 ft + 10 ft = 830 ft; therefore, the value of elevation for point E is 830 ft.

Figure 7.7

Apply these rules when trying to find the elevation:

1. If circular contour lines are between two contour lines that are not circular then the values of the circular contour line (starting from the outermost contour line) will have the same value of the highest non circular contour line:

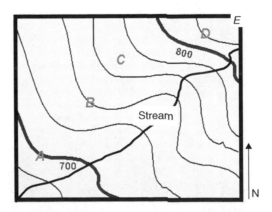

2. If circular contour lines are surrounded by circular contour lines then the values of the circular contour line (starting from the outermost contour line) will have value larger than the value of the circular contour line that is surrounding them:

3. If circular contour lines with hachures are between two contour lines that are not circular then the values of the circular contour line (starting from the outermost contour line) will have the same value of the lowest non circular contour line:

4. If circular contour lines with hachures are surrounded by circular contour lines, with no hachures, then the values of the circular contour line with hachures (starting from the outermost contour line) will have the same value of the circular contour line that is surrounding them:

5. If circular contour lines with hachures or without hachures are between two contour lines that are not circular then consider the noncircular contour lines first then the circular lines. The figure on left is wrong whereas the figure on the right is correct since the rule says always start with the noncircular lines first and then with the circular contour lines. For circular contour lines follow the rules mentioned above, in this case rule number (1).

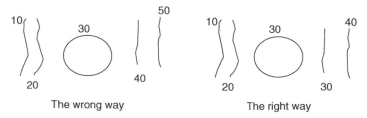

Gradient

Is the steepness of the slope, so high gradient reflects steep slope and low gradient reflects gentle slope. Look at Figure (7.8).

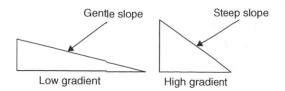

Figure 7.8

Mathematically, gradient is the vertical drop divided by horizontal distance so if you look at Figure (7.9) then G = (A - B)/X or (20 - 10 ft)/2 miles = 5 ft/mile

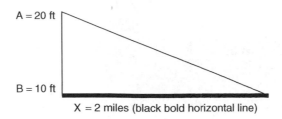

Figure 7.9

Gradient could be expressed using three methods:

1. Gradient could be expressed in degrees, measured down from the horizontal, so a vertical cliff has a gradient of 90°.
2. Gradient could be expressed in verbal form like saying 10 feet per mile and this was the definition mentioned earlier. See the previous example.
3. Gradient could be expressed as a ratio. For example 1:3 means that for every 3 horizontal units (any unit) there is a fall in height of 1 unit (same unit). Note that when gradient is expressed in this form, units are not mentioned or written like the fractional scale because they are cancelled.

Relief

Refers to the difference in elevation between any two points and there are maximum relief and local relief.

Maximum (total) relief

Is the difference between the highest point and the lowest point in the area being studied or considered.

Local relief

Is the difference between a high point and a low point in the local area of the map like a mountain and an adjacent valley.

Difference between Elevation, Height and Relief

Elevation is the distance taken from sea level to the top of an object. This is shown in figure (7.10) for object (A) and in this case object (A) is a mountain.

Height is the difference in distance between the top (head) of an object and the object's bottom (foot). This also is shown in Figure (7.10) and again object (A) is considered a mountain.

Relief is the distance difference between the highest point and lowest point in a specific (given) area. This is shown in Figure (7.10) and the specific (given) area is the dotted box.

Notice that relief is also used in a relative sense. For example, a mountain area has high relief whereas a plain area has low relief.

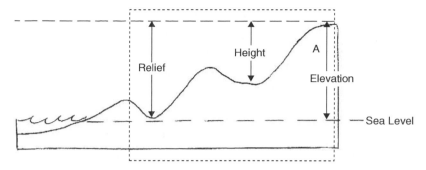

Figure 7.10

Public Land Survey System

To determine the location of a specific place you need to know the longitude and the latitude of that specific place.

Latitudes are imaginary horizontal lines that run from east to west and are also called parallels. Latitudes divide our planet to two halves; one called the northern hemisphere and the other one is called the southern hemisphere. Latitudes have values that start from 0° latitude to 90° latitude. The 0° latitude is called the equator.

Longitudes are imaginary vertical lines that run from north to south and are also called meridians. Longitudes divide our planet to two halves; one called the eastern half and the other one is called the western half. Longitudes have values that start from 0° longitude to 360° longitude. The 0° longitude is called prime meridian. The longitude opposite to 0° longitude, or 180° from the prime meridian is called the International Date Line. See Figure (7.11).

On topographic maps the longitude and the latitude are shown as black lines at the boundaries of the map and are written as black numbers at the corners of the topographic map, outside the map boundaries. Longitudes and latitudes are written as numbers that take the following form: 15° 30′ 00″ where ° represents degrees, ′ represents minutes, and ″ represents seconds.

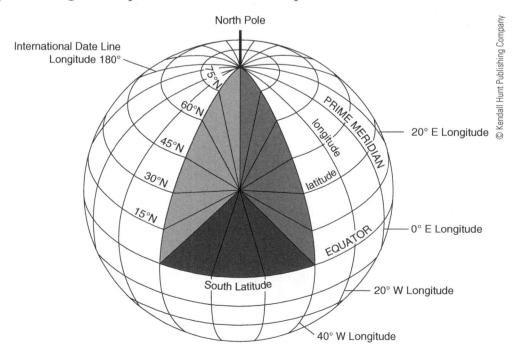

Figure 7.11

Note that latitudes and longitudes are written as degrees, minutes, and seconds because latitudes and longitudes represent angles. Also note that intersection of latitudes and longitudes gives deformed rectangles called quadrangles (four angles not equal to 90°) and the difference between two successive latitudes or two successive longitudes could be 7.5° (same as 7° 30′) or more than that. Look at Table (7.1).

In order to locate smaller areas we need a system that uses lines similar to the latitudes and longitudes and are parallel to them but cover smaller areas. This system is called Public Land Survey System (PLSS) and is used in 30 states and this surveying method is a way to subdivide and describe land in the United States. See Figure (7.12).

Like the longitude and latitude system the PLSS is made of squares produced by intersection of horizontal lines parallel to initial horizontal line called Base Line and vertical lines parallel to initial vertical line called Principal Meridian.

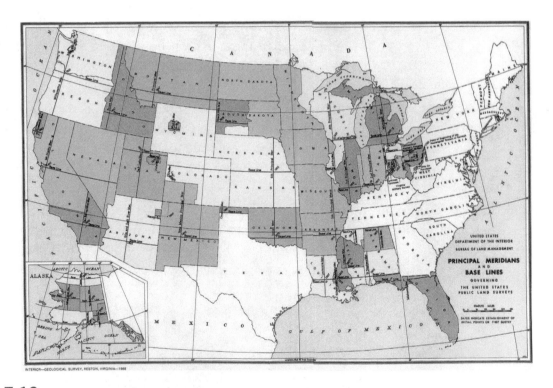

Figure 7.12

Source: Principal Meridians and Base Lines, Bureau of Land Management

The lines above and below the Base Line will be denoted Township. Lines above Base Line will be North with increasing numbers from Base Line and lines below Base Line will be South with increasing numbers from Base Line.

Same thing with Principal Meridian, any line to the right or left of Principal Meridian will be denoted Range. Lines to right of the Principal Meridian will be East with increasing numbers from the Principal Meridian and lines to the left of the Principal Meridian will be West with increasing numbers from the Principal Meridian. See Figure (7.13).

Each specific certain states have their own base line and prime meridian. Look at Figure (7.12).

The Township is written as: Township Number North or South.

For example: T3N.

The Range is written as: Range Number East or West.

For example: R1E.

Both Township and Range are written as above forms in red outside the topographic map.

Each Township is 36 miles^2.

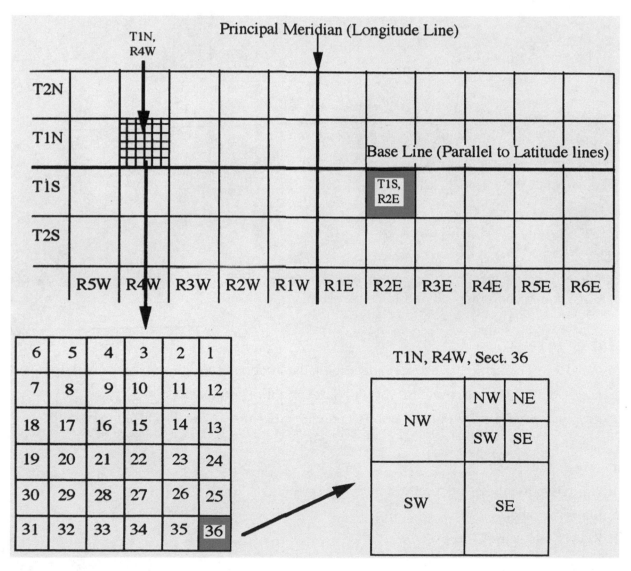

Figure 7.13

Furthermore, each Township is divided to 36 sections. See Figure (7.13). The beginning of sections within Township starts at the upper right corner of the Township then numbers increase as illustrated.

Each section is 1 mile^2 and is equal to 640 acres. Sections are written in red as numbers inside the topographic map.

Notice that sections could be divided and described as shown in Figure (7.14).

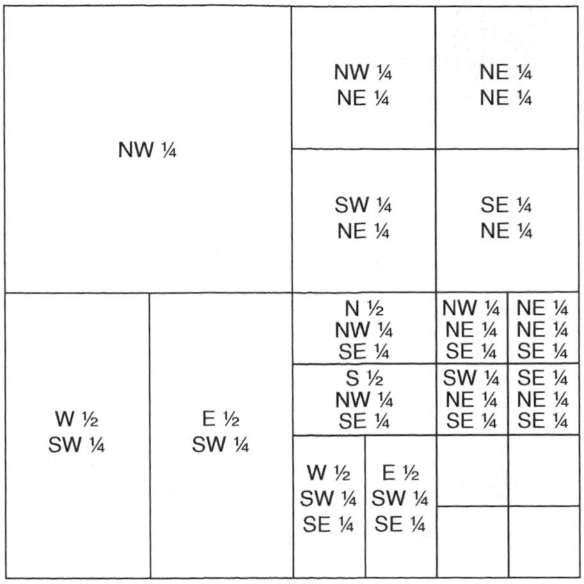

Typical section subdivisions

Figure 7.14

When describing the location of a specific place using the PLSS it is advised to start from township then range then section then subdivision of a section.

Example

Using the PLSS and using Figure (7.15) determine the location of the thick black boundary square.

The answer is T2S R3W Section 36 NW1/4 NE1/4 and it is read like this:

Northeast quarter of northwest quarter of section 36, range three west, township two south.

Remember:

1 township is = 36 miles^2 and 1 township is = 36 sections

1 section is = 1 mile^2

1 section is = 640 acres

Look at figure (7.13) and figure (7.15).

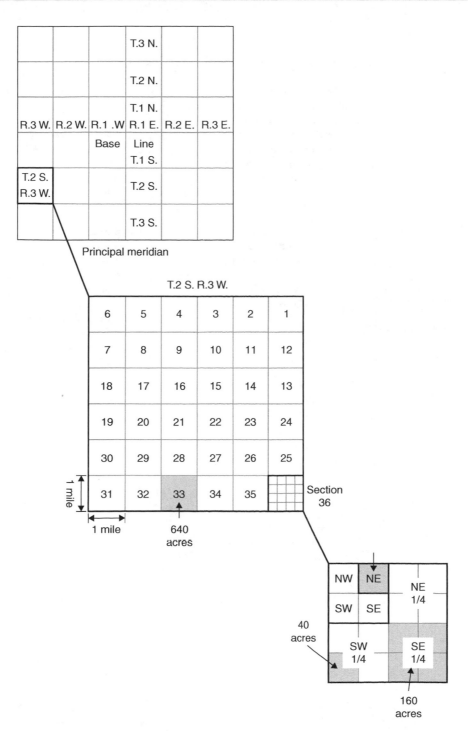

Figure 7.15

Some of the symbols used in topographic maps:

Topographic Map Symbols

BOUNDARIES
National
State or territorial
County or equivalent
Civil township or equivalent
Incorporated-city or equivalent
Park, reservation, or monument
Small park

LAND SURVEY SYSTEMS
U.S. Public Land Survey System:
Township or range line
Location doubtful
Section line
Location doubtful
Found section corner; found closing corner
Witness corner; meander corner

Other land surveys:
Township or range line
Section line
Land grant or mining claim; monument
Fence line

ROADS AND RELATED FEATURES
Primary highway
Secondary highway
Light duty road
Unimproved road
Trail
Dual highway
Dual highway with median strip
Road under construction
Underpass; overpass
Bridge
Drawbridge
Tunnel

BUILDINGS AND RELATED FEATURES
Dwelling or place of employment: small; large
School; church
Barn, warehouse, etc.: small; large
House omission tint
Racetrack
Airport
Landing strip
Well (other than water); windmill
Water tank: small; large
Other tank: small; large
Covered reservoir
Gaging station
Landmark object
Campground; picnic area
Cemetery: small; large

RAILROADS AND RELATED FEATURES
Standard gauge single track; station
Standard gauge multiple track
Abandoned
Under construction
Narrow gauge single track
Narrow gauge multiple track
Railroad in street
Juxtaposition
Roundhouse and turntable

TRANSMISSION LINES AND PIPELINES
Power transmission line: pole; tower
Telephone or telegraph line
Aboveground oil or gas pipeline
Underground oil or gas pipeline

CONTOURS
Topographic:
Intermediate
Index
Supplementary
Depression
Cut; fill
Bathymetric:
Intermediate
Index
Primary
Index Primary
Supplementary

MINES AND CAVES
Quarry or open pit mine
Gravel, sand, clay, or borrow pit
Mine tunnel or cave entrance
Prospect; mine shaft
Mine dump
Tailings

SURFACE FEATURES
Levee
Sand or mud area, dunes, or shifting sand
Intricate surface area
Gravel beach or glacial moraine
Tailings pond

VEGETATION
Woods
Scrub
Orchard
Vineyard
Mangrove

COASTAL FEATURES
Foreshore flat
Rock or coral reef
Rock bare or awash
Group of rocks bare or awash
Exposed wreck
Depth curve; sounding
Breakwater, pier, jetty, or wharf
Seawall

BATHYMETRIC FEATURES
Area exposed at mean low tide; sounding datum
Channel
Offshore oil or gas: well; platform
Sunken rock

RIVERS, LAKES, AND CANALS
Intermittent stream
Intermittent river
Disappearing stream
Perennial stream
Perennial river
Small falls; small rapids
Large falls; large rapids
Masonry dam
Dam with lock
Dam carrying road
Intermittent lake or pond
Dry lake
Narrow wash
Wide wash
Canal, flume, or aqueduct with lock
Elevated aqueduct, flume, or conduit
Aqueduct tunnel
Water well; spring or seep

GLACIERS AND PERMANENT SNOWFIELDS
Contours and limits
Form lines

SUBMERGED AREAS AND BOGS
Marsh or swamp
Submerged marsh or swamp
Wooded marsh or swamp
Submerged wooded marsh or swamp
Rice field
Land subject to inundation

Figure 7.16

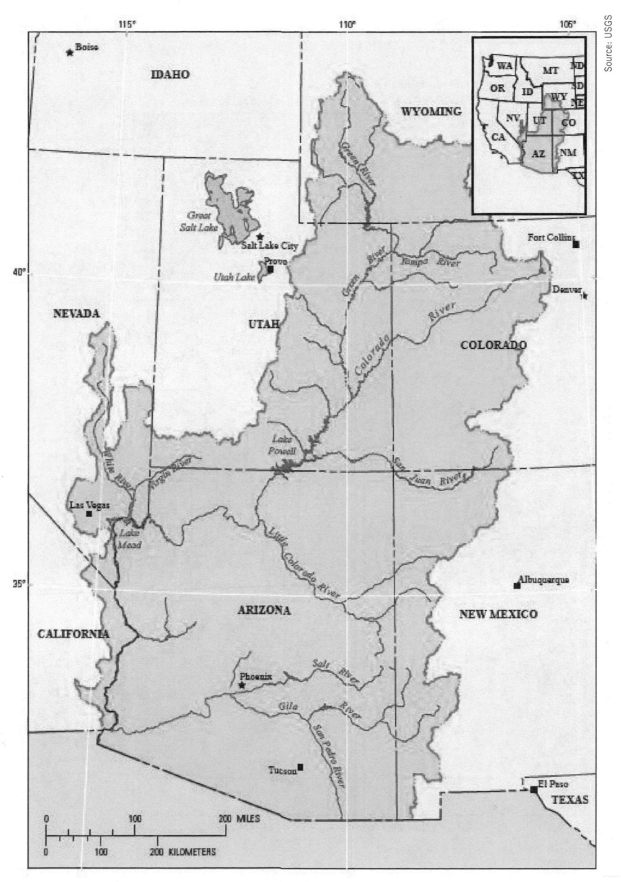

Source: USGS

1: This map shows Green River and Colorado River and path of flowing

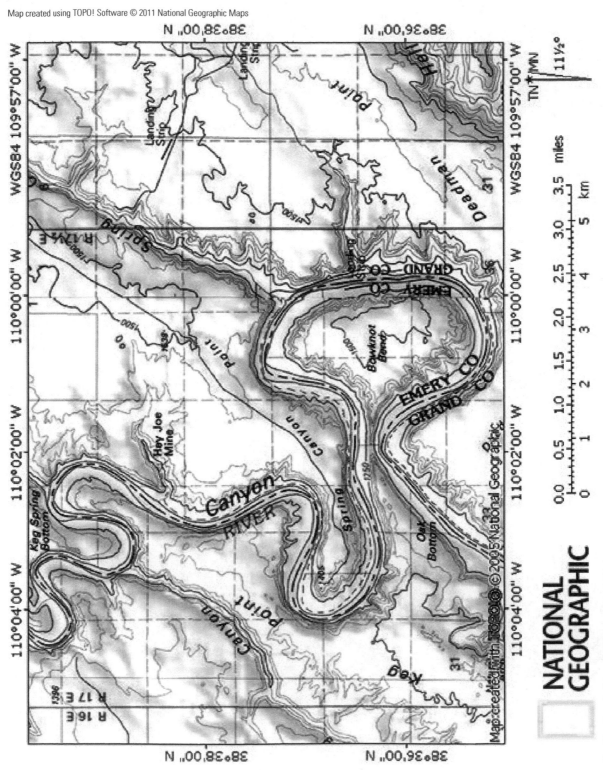

Map created using TOPO! Software © 2011 National Geographic Maps

NATIONAL
GEOGRAPHIC

2: Part of Green River in Utah

Figure 8.9

PART (E)

Figure (8.10) is showing part of Catskill Mountains (northeastern portion) in New York. The Catskill Mountains are part of northeastern part of the Appalachian Mountains.

The western part of the map is underlain by sedimentary rocks with nearly horizontal layers and the Schoharie Creek is passing through it, whereas the east part is underlain by steeply tilted sedimentary layers and Plattekill Creek is passing through it.

1.E) There are two drainage basin types shown in this map. One type is dendritic and the other type is rectangular. Try to determine the type of drainage basins for the east side of the map and the west side of the map.

2.E) Is Schoharie Creek passing through steep or gentle slope?

3.E) Is Plattekill Creek passing through steep or gentle slope?

4.E) Based on your answers to questions (2.E) and (3.E) which stream out of the two will erode towards the head (headword erosion) and why?

5.E) Based on your answer to question (4.E) what will be the fate of Schoharie Creek (to be more specific the tributaries of Schoharie Creek)?

6.E) Your answer to question (5.E) is referring to a geologic process that happens between streams. What is the name of this geologic process?

7.E) Rocks on both sides of the map are sedimentary rocks. Which part (side) of the map, eastern or western is more resistant to erosion? Why?

Map created using TOPO! Software © 2011 National Geographic Maps

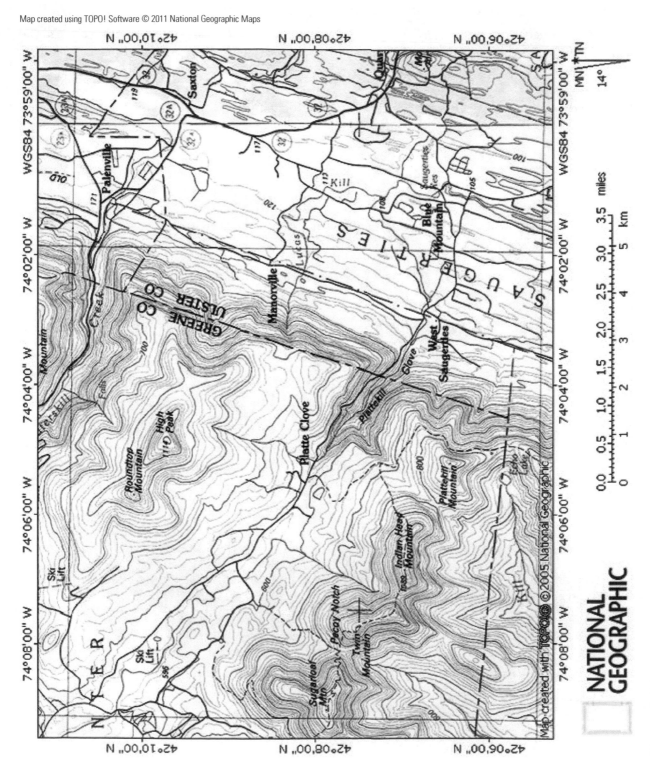

NATIONAL
GEOGRAPHIC

Figure 8.10

PART (F)

This exercise will guide you to draw a drainage basin using the following guide lines:

1. Try to locate the mouth (point of discharge) of the specific stream. If the mouth of the stream is not in the map or outside the range of the map then locate the lowest elevation of stream and remember streams flow from high elevation to low elevation. Usually the mouth of the stream is at another stream (intersection of two streams) or lake or ocean (sea).

2. Using trace paper on the topographic map, try to trace the streams or creeks and the branching streams that are shown on the map. The branching streams are feeding the specific major stream (flowing in to the main stream). Remember streams are shown as solid blue lines or broken blue lines. Now you defined the primary drainage network.

3. Add additional tributaries that are not seen on the map to the tracing paper. You can do that by noticing the contour lines that make a "v" shape, that is, whenever there is a v-shaped contour line then it is possible to draw a line through it and that will represent a branching stream that is draining in the main stream. Remember that these additional streams are not shown on the map and you added them to the tracing paper. Furthermore when you select a v-shaped contour line it must be related to the specific major stream and the v-shaped contour line is the one that indicates the direction of the branching stream that drains in the major specific stream. Furthermore, remember that the closing of the v-shaped contour line points to the uphill. This step is difficult to follow since some of the v-shaped contour lines are not the required ones. For example, those v-shaped contour lines that defy the logic that is they indicate flow of streams is uphill and not downhill then these should not be considered.

4. Define adjacent drainage basin. This step is neglected if there is no other major stream in the topographic map.

5. If there is another major stream in the map then repeat steps (1)–(3). Use another color for this second drainage network (basin).

6. Locate on the topographic map the places of top hills, top mountains, and ridges by putting dots. To do that find out contour lines that make circles or nearly complete circles since they represent top of hills or top of mountains and ridges. Make sure that these points are separating the specific drainage network from the other drainage networks. Shade these circles or put dots on these places that represent top hills, top mountains and ridges.

7. Define the divide of drainage basin. Simply connect the above dots (obtained from step 6) for the specific major drainage network (not the other adjacent network) and this will produce the boundary (divide) of the major drainage network and hence the drainage basin. The divide should contain all the tributaries of the specific major basin and should not cross any of the tributaries so the divide is a boundary between the tributaries of the major drainage network and other tributaries that are not part of the major drainage network.

8. If your work is correct then all the water (or assume a droplet) that falls in to the specific drainage basin will flow inside the basin and any water that flows outside the basin is not considered part of the specific drainage basin.

Using the previous guidelines try to outline the drainage network to draw drainage basin on figure (8.11).

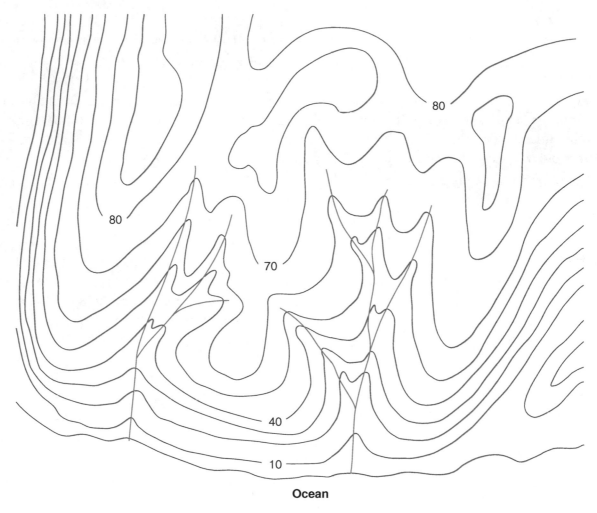

Figure 8.11

PART (G)
‖‖

The following example (table, numbers, and information) was taken from USGS website: (http://ga.water.usgs.gov/edu/streamflowpatterns.html).

Table (8.4) shows the annual mean stream flow at Peachtree Creek at Atlanta for the years 1959–2001. Annual mean stream flow is the average of all daily stream flows for the year.

1.G)   Find the magnitude (m) by putting discharge in order starting from the highest as 1 in Table (8.4). Some values are already given in Table (8.4).

2.G)   Calculate the probability of recurrence as percentage for each year using the formula:

$$P = (m/(n+1)) * 100$$

where m is the magnitude (rank or order of discharge) and n is the number of years and in this case is 43 years.

3.G)   Plot discharge against probability from Table (8.4) on the graph paper supplied, then draw a best fit straight line through the points to construct the flood-frequency curve.

4.G)   Find the recurrence interval for discharge equal to 70 ft^3/s. You can use the flood-frequency curve obtained in question (3.G) from the best fit straight line or using the following formula:

$$R.I. = (n+1)/m$$

where n is the number of years and in this case is 43 years and m is the magnitude (rank or order of discharge).

It is important to mention that for this exercise it is possible to replace discharge in Table (8.4) with maximum stage where maximum stage is maximum elevation of water in channel or maximum amount of water a channel can hold. Furthermore, stage, is referring to elevation of water in stream taken from a reference point.

Probability (P) is the chance that a stream will reach maximum stage (flood), or discharge in any given year so if for example, one stream has probability equal to 20% and another stream has probability equal to 70% that means the second stream ($P = 70\%$) will have more chance to flood in a specific year compared to the first stream ($P = 20\%$) or the chance for flooding (reaching maximum stage) for the second stream is more than the first stream in a specific year.

Also notice that recurrence interval (R.I.) is referring to the chance that a stream will flood (reach maximum stage), at least once, every specific number of years, so for example if R.I. is equal to 50 years that means the specific stream will have chance to flood at least once every 50 years.

Table 8.4

Year	Annual Mean Stream Flow (Discharge) in ft³/s	Magnitude, m	Probability Recurrence, P (as percentage)
1959	78.4		
1960	87.9		
1961	138		
1962	112		
1963	137		
1964	188	2	
1965	102		
1966	155		
1967	172		
1968	134		
1969	145		
1970	119		
1971	137		
1972	134		
1973	179	3	
1974	143		
1975	213	1	
1976	148		
1977	118		
1978	97.7		
1979	163		
1980	149		
1981	70.0		
1982	145		
1983	165		
1984	172		
1985	125		
1986	82.4		
1987	116		
1988	81.2		
1989	166		
1990	175		
1991	175		
1992	171		
1993	125		
1994	150		
1995	157		
1996	153		
1997	155		
1998	139		
1999	86.9		
2000	91.4		
2001	99.8		

GROUNDWATER, KARST TOPOGRAPHY AND STREAMS

1. Figure (9.1) shows map of the USA and Canada where the average annual precipitation is represented by different colors. See the legend. Areas with more than 50 cm per year are considered humid.

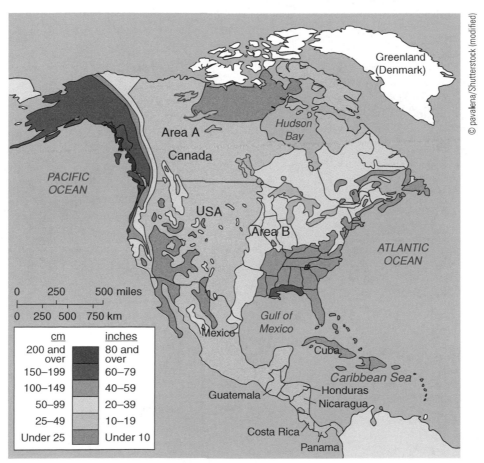

Figure 9.1

2. Gaining streams usually occur in humid regions and the water table is closer to surface; therefore, it supplies water to streams. In arid regions, the water table is not close to the surface as in humid regions; therefore, the stream supplies water to ground water and these streams are called losing streams.

3. Look at Figure (9.2) and study it carefully. Note that the solid brown lines are topographic contour lines whereas blue circular lines with numbers inside them represent small lakes. Also note that there is a stream crossing the topographic contour lines.

Water table contour lines are lines that connect point of equal elevation of water table from sea level. The numbers in lakes in Figure (9.2) represent the elevation of water table.

Construct water table contour lines by connecting points of equal elevation of water table, using the lakes. Draw water table contour lines on Figure (9.2). These lines (water table contour lines) stop at the stream from both sides of the stream and they don't cross the stream to the other side.

Flow lines are lines that show the flow of ground water. These lines are drawn from high water table contour lines to low water table contour lines perpendicular to water table contour lines. Flow lines converge or diverge but never intersect or cross each other, that is, flow lines on either side of stream will not cross to the other side of stream. Draw flow lines on Figure (9.2).

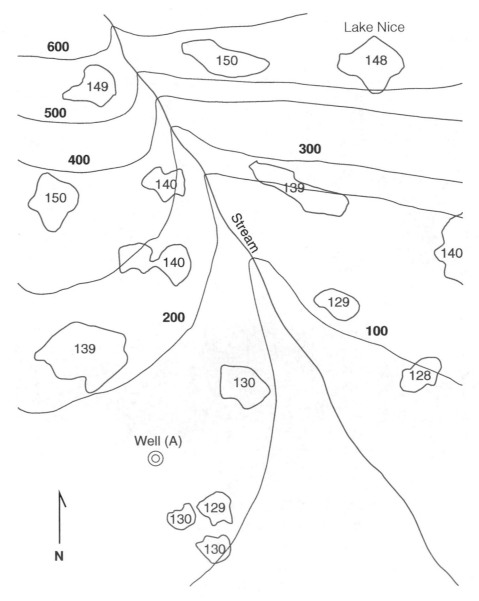

Figure 9.2

4. What direction is the stream flowing?

   a) NE to SW   b) NW to SE   c) SW to NE

5. What is the direction of flow of ground water?

   a) Flowing same direction of stream   b) Flowing opposite direction of stream

6. Is the ground water flow lines converging toward the stream or diverging away from the stream?

   a) Converging towards the stream   b) Diverging away from the stream

7. Is this stream losing or gaining stream?

   a) Losing stream   b) Gaining stream

8. Does the area shown in Figure (9.2) represent dry or humid region?

   a) Dry   b) Humid

9. Based on your answer to question (8) and using Figure (9.1), this region is found in:

   a) Area (A)   b) Area (B)

10. The answer for question (8) suggests that the stream in Figure (9.2) is:

   a) Losing stream   b) Gaining stream

11. From your answers to question (6) and question (8), it is possible to say that there is relationship between type of climate and type of stream (losing or gaining stream).

   a) True   b) False

12. If polluted material were dumped in Lake Nice, will the stream be polluted?

   a) Yes   b) No

13. If your answer to question (12) is "Yes," then why is it that the stream will be polluted?

   a) Ground water is supplying stream with water

   b) Stream is supplying water to ground water

14. Will well (A) be polluted?

   a) Yes   b) No

15. If the answer to question (14) is "Yes," then this answer is based on that:

   a) Ground water flows from areas of high water table contour lines to low water table contour lines

   b) The polluted water of the stream will pollute the ground water that supplies water to well (A)

16. At well (A), how deep you need to dig in order to get to water table?

   a) 135 ft   b) 150 ft   c) 15 ft   d) 215   e) 10 ft

17. Study Figure (9.3). This figure shows different types of drainage basin patterns.

   Now look at Figure (9.4). It shows a block diagram. The top of the block diagram represents the surface whereas the sides of the block diagram represent the cross sections that show different types of rocks, layers and structures like fractures, faults, and magma intrusion.

   See the legend for explanation.

   Note that there are no streams on the surface.

   If the surface shows no streams, as mentioned earlier, what type of drainage pattern would develop at locations (A) in the future?

   a) Dendritic   b) Rectangular   c) Trellis   d) Radial

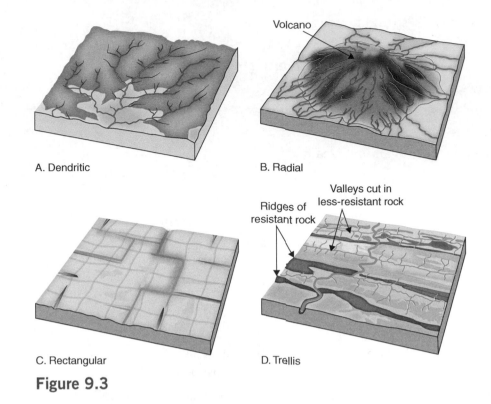

A. Dendritic

B. Radial

Valleys cut in
less-resistant rock

Ridges of
resistant rock

C. Rectangular

D. Trellis

Figure 9.3

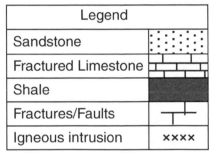

Legend	
Sandstone	
Fractured Limestone	
Shale	
Fractures/Faults	
Igneous intrusion	××××

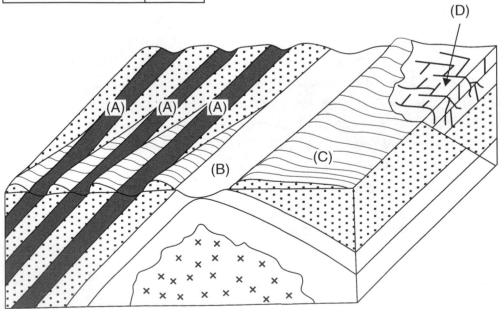

Figure 9.4

18. The reason for your answer to question (17) is because:
    a) There are alternating layers of resistant and less resistant rocks.
    b) There are fractures and faults in specific type of rocks.
    c) There is magma activity that could produce volcanic activity (volcano).
    d) There are relatively uniform rocks (resistance to weathering and erosion is uniform) like sandstone, igneous rocks (granite), or even sediments.

19. What type of drainage pattern would develop at location (B) in the future?
    a) Dendritic   b) Rectangular   c) Trellis   d) Radial

20. The reason for your answer to question (19) is because:
    a) There are alternating layers of resistant and less resistant rocks.
    b) There are fractures and faults in specific type of rocks.
    c) There is magma activity that could produce volcanic activity (volcano).
    d) There are relatively uniform rocks (resistance to weathering and erosion is uniform) like sandstone, igneous rocks (granite) or even sediments.

21. What type of drainage pattern would develop at location (C) in the future?
    a) Dendritic   b) Rectangular   c) Trellis   d) Radial

22. The reason for your answer to question (21) is because:
    a) There are alternating layers of resistant and less resistant rocks.
    b) There are fractures and faults in specific type of rocks.
    c) There is magma activity that could produce volcanic activity (volcano).
    d) There are relatively uniform rocks (resistance to weathering and erosion is uniform) like sandstone, igneous rocks (granite), or even sediments.

23. What type of drainage pattern would develop at location (D) in the future?
    a) Dendritic   b) Rectangular   c) Trellis   d) Radial

24. The reason for your answer to question (23) is because:
    a) There are alternating layers of resistant and less resistant rocks.
    b) There are fractures and faults in specific type of rocks.
    c) There is magma activity that could produce volcanic activity (volcano).
    d) There are relatively uniform rocks (resistance to weathering and erosion is uniform) like sandstone, igneous rocks (granite), or even sediments.

25. Look at Figure (9.5). It shows drainage pattern developed on Mars that could be thousands or millions of years old. What type of drainage pattern is it?
    a) Dendritic   b) Rectangular   c) Trellis   d) Radial

26. Can you conclude any type of information from Figure (9.5) based on your answer to question (25)?
    a) There are alternating layers of resistant and less resistant rocks.
    b) There are fractures and faults in specific type of rocks.
    c) There is magma activity that could produce volcanic activity (volcano).
    d) There are relatively uniform rocks (resistance to weathering and erosion is uniform) like sandstone, igneous rocks (granite), or even sediments.

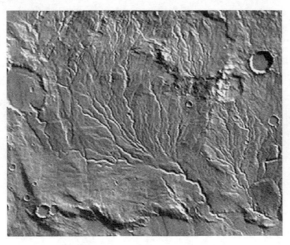

NASA/JPL/Malin Space Science Systems

Figure 9.5

27. Figure (9.6) shows the development of Karst topography through three stages and the features associated with each stage. Study Figure (9.6) carefully then answer the following questions.

    West Texas like Manati in Puerto Rico shows karst topography; however, West Texas shows different stage of development than Manati in Puerto Rico. In your opinion, which one of the two locations should show more developed stage of Karst topography?

    a) West Texas   b) Manati in Puerto Rico

28. Why is it that West Texas and Manati in Puerto Rico show different stages of Karst topography?

    a) The type of rocks are different   b) The climate is different

29. Which one out of the two locations (West Texas and Manati in Puerto Rico) is humid?

    a) West Texas   b) Manati in Puerto Rico

30. Is there a relationship between climate and weathering (chemical weathering)?

    a) Yes   b) No

31. Is there a relationship between climate and stage of development of Karst topography?

    a) Yes   b) No

32. What is the name of those isolated erosional remnants found in the late stage (C)?

    a) Solution valleys   b) Disappearing streams

    c) Sinks   d) Towers

    Look at Figure (9.7) and study it carefully then select the correct answer for the following questions.

33. Feature (A) is/are:

    a) Sinkhole(s)   b) Solution valley

    c) Towers   d) Limestone

    e) Caves   f) Disappearing or sinking stream (creek)

34. Feature (D) is/are:

    a) Sinkhole(s)   b) Solution valley

    c) Towers   d) Limestone

    e) Caves   f) Disappearing or sinking stream (creek)

A. Early Stage:
(1) Surface is nearly flat with few small, scattered sinkhole depressions. Subterranean caverns are numerous.
(2) Throughout the early stage, sinkholes become more abundant and increase in size. Sinking streams could develop at this stage.

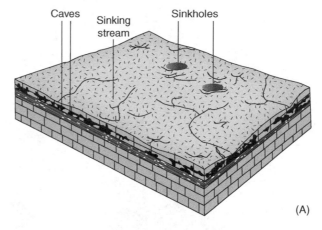

B. Intermediate Stage:
(1) Individual sinkholes enlarge and merge to form solution valleys with irregular branching outlines.
(2) Much of the original surface is destroyed. There are many springs and disappearing streams.
(3) Maximum relief, although not great, is achieved. Differences in elevation between the rim and floor of a sinkhole rarely exceed 200 to 300 ft.

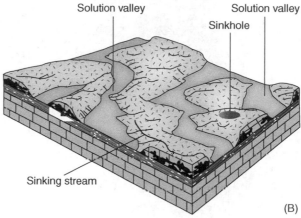

C. Late Stage:
(1) Solution activity has reduced the area to the base of the limestone unit.
(2) Hills formed as erosional remnants are few, widely scattered, and generally reduced to low, conical knots known as Towers.

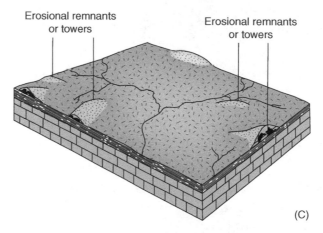

Figure 9.6

35. Feature (E) is/are:
   a) Sinkhole(s)
   b) Solution valley
   c) Towers
   d) Limestone
   e) Caves
   f) Disappearing or sinking stream (creek)
36. Feature (F) is/are:
   a) Sinkhole(s)
   b) Solution valley
   c) Towers
   d) Limestone
   e) Caves
   f) Disappearing or sinking stream (creek)

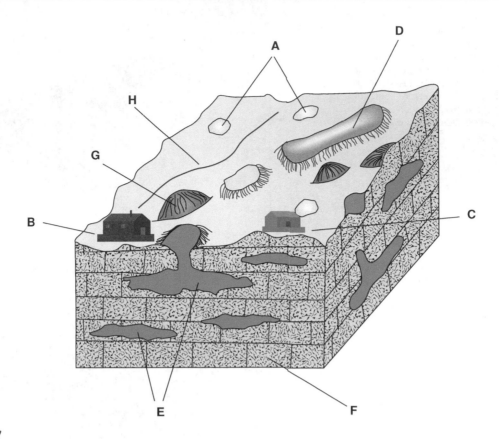

Figure 9.7

37. Feature (G) is/are:

   a) Sinkhole(s)   b) Solution valley

   c) Towers   d) Limestone

   e) Caves   f) Disappearing or sinking stream (creek)

38. Feature (H) is/are:

   a) Sinkhole(s)   b) Solution valley

   c) Towers   d) Limestone

   e) Caves   f) Disappearing or sinking stream (creek)

39. There are two houses one at location (B) and the other house at location (C). Which one of the two houses will be damaged with time?

   a) House at location (B)   b) House at location (C)

40. Does the Karst topography shown in Figure (9.7) indicate an early stage of development or inter-mediate/late stage of development?

   a) Early stage of development   b) Intermediate/late stage of development

41. The Karst topography shown in Figure (9.7) is an indication that the climate was (humid, arid) and that the rocks were (soluble, insoluble). Select the correct answer.

   a) Arid, insoluble   b) Humid, insoluble

   c) Arid, soluble   d) Humid, soluble

Figure (9.8) shows the topographic map for the northern part of Florida close to Wellborn. The area east to Wellborn is characterized by sink holes. Answer the following questions after you study the map carefully.

Map created using TOPO! Software © 2011 National Geographic Maps

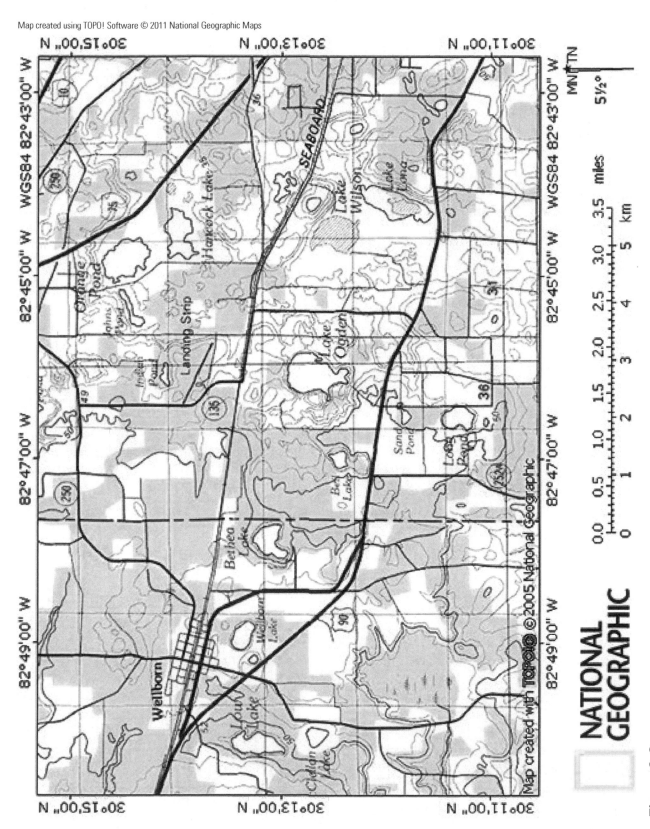

NATIONAL
GEOGRAPHIC

Figure 9.8

42. What is the origin of the lakes and ponds seen in the area that is how did they form?

43. Are there any depression(s) in the area covered by the map?

44. How do depressions look like in a topographic map that is how depressions are shown on a topographic map?

45. Florida is underlain by a specific type of rock that is causing these depressions to form. What is the name of that rock?

46. Look carefully at the map and you will notice that some depressions are not filled with water. Why is it that there is no water in some of these depressions?

47. Is the water table close to the surface? Answer with yes or no and then support your answer with evidence(s).

48. Look at Figures (9.9) and (9.10). Figure (9.9) shows portion of topographic map of Royal Gorge, Colorado, where Arkansas River passes through the gorge. Figure (9.10) shows portion of topographic map for Pearl River, Mississippi/Louisiana. Compare the two maps and specifically the two streams. Notice the features that are associated with the two streams then answer the following questions.

    The flood plain of Pearl River is _____ (than) the flood plain of Arkansas River.

    a) Wider   b) Narrower   c) Same as

49. Erosion produced by Pearl River is _____ whereas erosion produced by Arkansas River is

    _____.

    a) Lateral, down cutting   b) Down cutting, lateral

50. Which one of the two streams has more of the following features: oxbow lakes, back swamps, more meanders and cutoffs.

    a) Arkansas River   b) Pearl River   c) Both streams

51. Pearl River should show _____ levees than Arkansas River.

    a) More   b) Less

52. Which one of the two streams shows early stage of development?

    a) Arkansas River   b) Pearl River

53. Which one of the two streams shows late stage of development?

    a) Arkansas River   b) Pearl River

Map created using TOPO! Software © 2011 National Geographic Maps

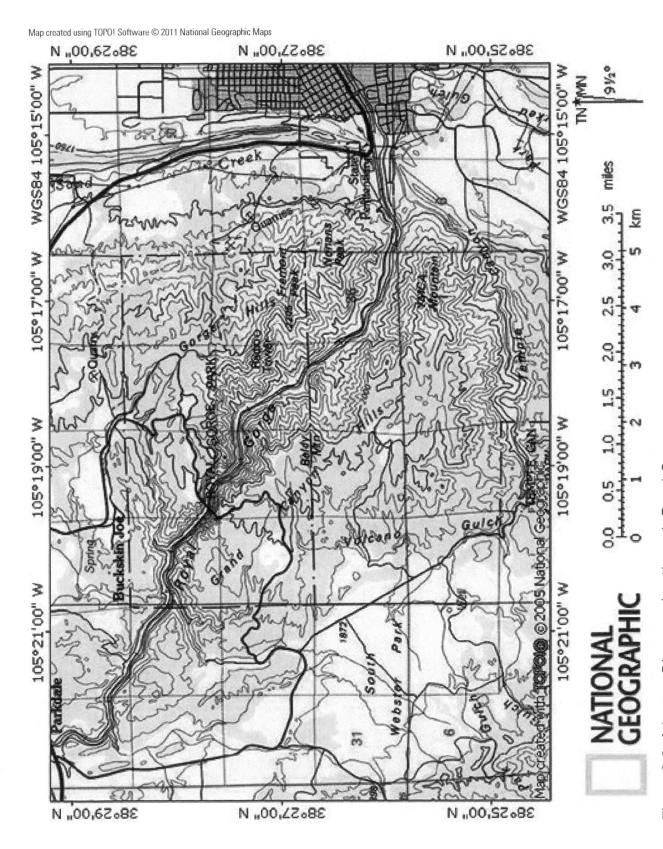

Figure 9.9 Arkansas River passing through Royal Gorge

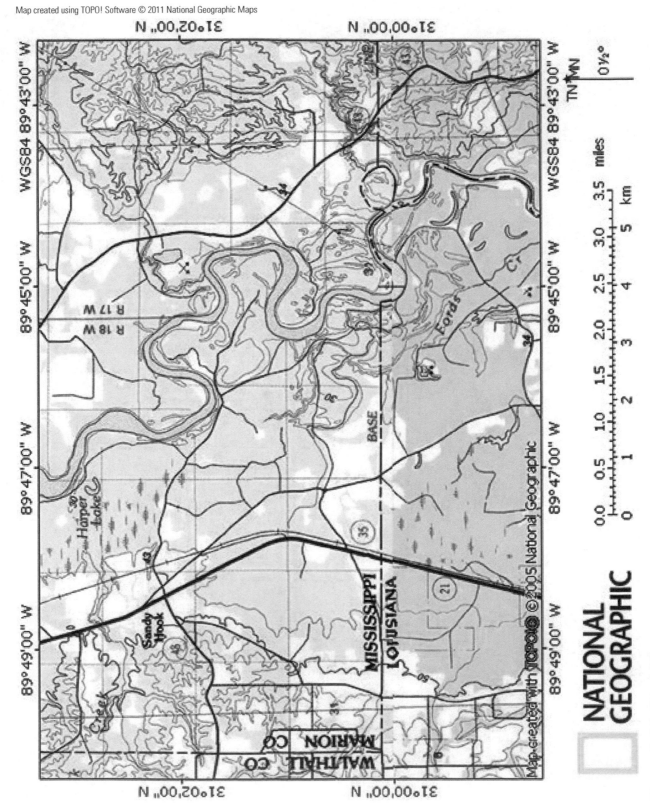

Figure 9.10 Pearl River

The development of Karst Topography and the geological features related to it has negative influence on structures and human life, and therefore, building of any structure on the surface should be regulated not only based on engineering and civil codes but also on geological concerns and concepts.

Interestingly the influence of karst topography features like sinkholes has its effect on health. For example 40% of groundwater used for drinking in the USA comes from karst aquifers (cavities formed in soluble rocks), and it is known that 20% of land surface in the USA is karst. Figure (9.11) shows the map of the USA where evaporite rocks like saltrock (halite), gyprock (gypsum), and anhydrite and carbonate rocks like limestone and dolomite underlie the USA. The evaporites and carbonate rocks are soluble rocks and they increase the rate of karst topography development.

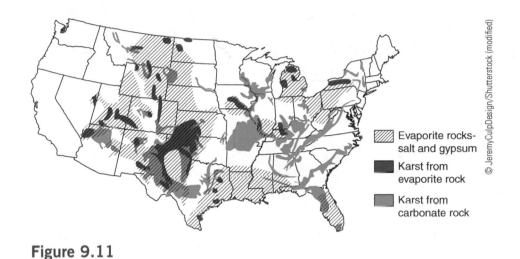

Figure 9.11

Figure (9.12) shows soluble rocks like limestone and an aquifer developed in the limestone. Furthermore the figure shows structures on the surface and other features. Study the figure carefully and then answer the following questions based on the knowledge you gained from this exercise, and the information provided earlier.

54. Can the aquifer developed in the limestone also be considered a cave (cavern)?

55. What are the sources of pollution in Figure (9.12)? List them.

56. Is limestone like sandstone or shale, a good filter for pollutants?

57. What are some measures and regulations that should be taken and considered before building any structure on the surface?

58. What are some of the problems that will develop in the future and are related to the situation in Figure (9.12)?

59. Mention some of the practical solutions that could be used or practiced to avoid problems that will rise in the situation of Figure (9.12).

60. Study Figure (9.13) carefully and then mention the factors that could attribute to karst topography development.

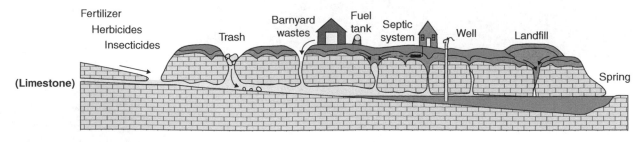

Figure 9.12

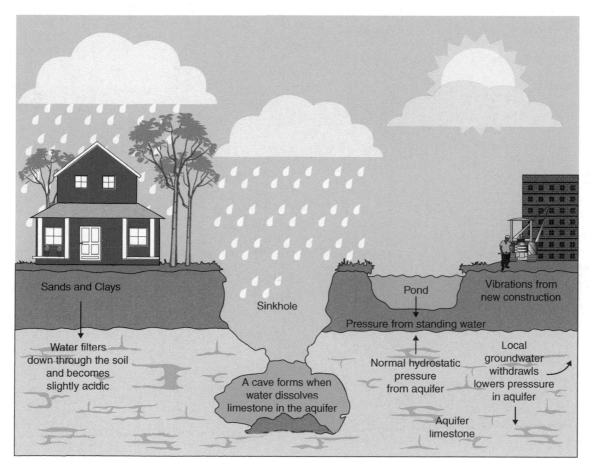

Figure 9.13 The Making of a Sinkhole

Map created using TOPO! Software © 2011 National Geographic Maps

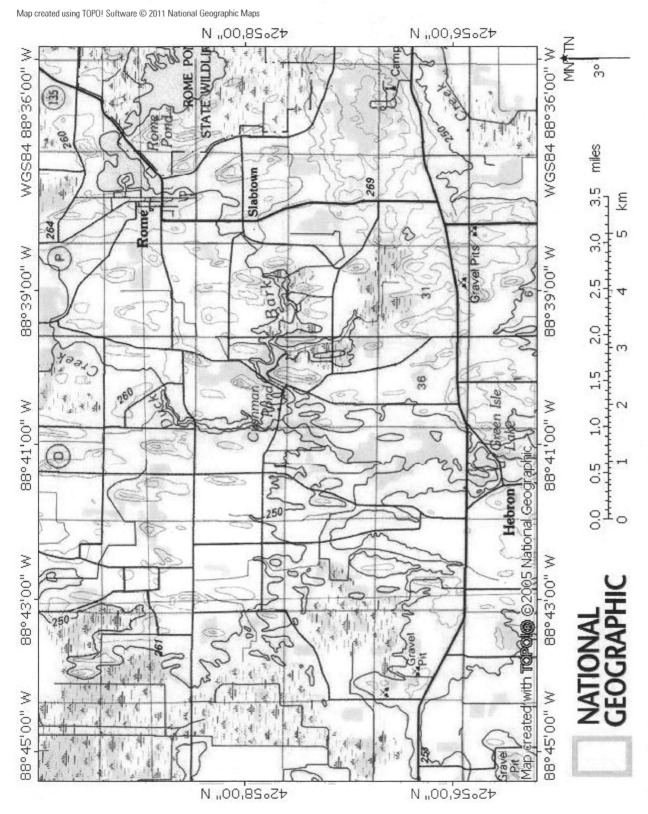

NATIONAL
GEOGRAPHIC

Figure 10.6

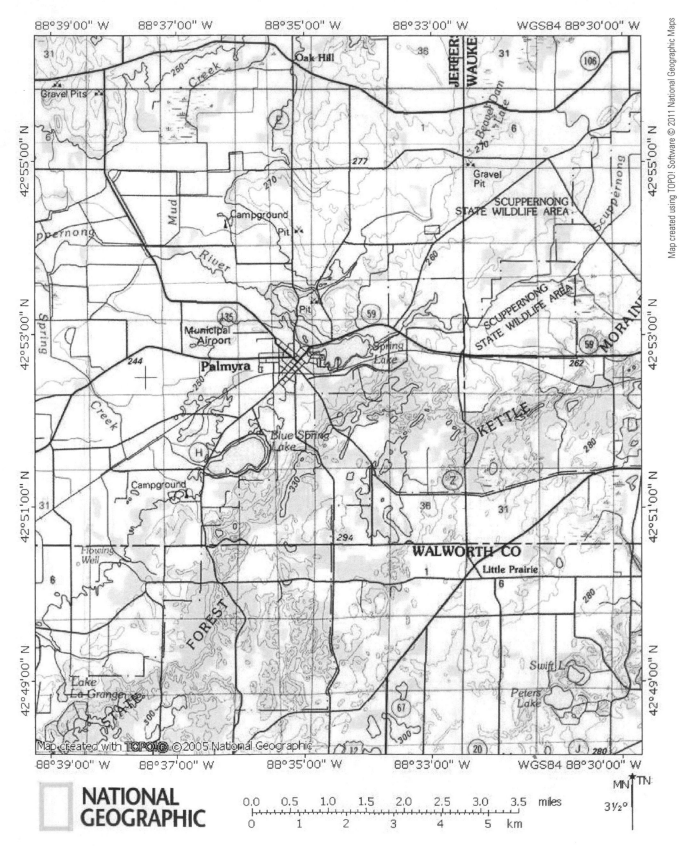

NATIONAL
GEOGRAPHIC

Figure 10.7

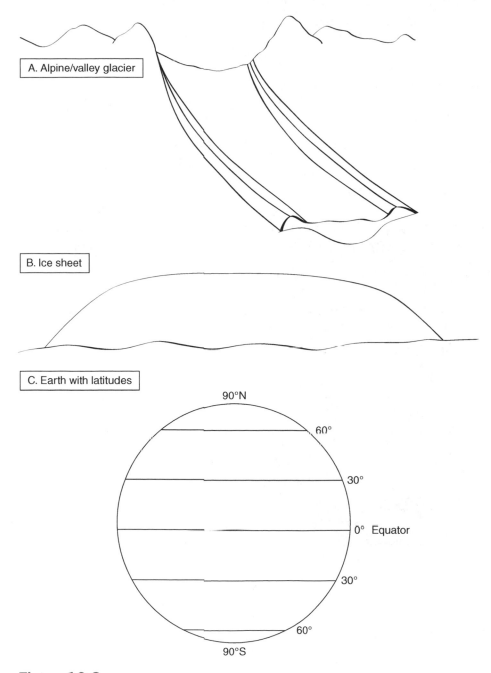

A. Alpine/valley glacier

B. Ice sheet

C. Earth with latitudes

90°N

60°

30°

0° Equator

30°

60°

90°S

Figure 10.8

PART (F)

Figure (10.8) is showing three parts. Part (A) shows simplified drawing of alpine/valley glacier with lateral moraines on sides; Part (B) shows ice sheet glacier, and part (C) shows planet earth with latitudes.

1.F)   What is meant by "budget" of glacier?

2.F)   List the components of glacier's budget.

3.F)   What is the definition of snow line?

4.F)   Label zone of accumulation and zone of ablation for both the alpine/valley glacier (part A) and ice sheet (part B).

5.F)   After answering question (4.F) draw snow line for both alpine/valley glacier (part A) and ice sheet glacier (part B). Show both snow lines on the two type glaciers using the same one color.

6.F)   Using the knowledge that you obtained from the previous answers and using part (C) of Figure (10.8), try to find a relation between snow line on alpine/valley glacier and latitude. State that relationship.

VOLCANOES

PART (A)

1.A) Given the information about the different types of volcanoes and volcanic activity and referring to Figure (11.1), try to match the drawings (numbers) in the figure with the specific term describing type of volcano and volcanic activity.

Cinder Cone

Cinder Volcanoes are the simplest type of volcano. They are built from rock fragments produced by volcanic eruption (pyroclasts). Although this type of volcanoes is made of pyroclasts however, sometimes, lava will flow from the base of the volcano. Most cinder cones have a bowl-shaped crater at the summit, and this crater is usually deep, and rarely rise more than a thousand feet or so above their surroundings.

Cinder cones are smaller than shield and composite volcanoes but larger than volcanic domes and have steep slopes. They are either produced from basaltic magma rich in gases or magma of acidic composition. These volcanoes are found at sides of Mt. Etna and also in Arizona, in Flagstaff area.

Composite Volcanoes

Typically, composite volcanoes are steep-sided, symmetrical cones of large dimension built of alternating layers of lava flow, volcanic ash and cinders (pyroclasts). Composite volcanoes will rise as much as 8000 feet (2438.4 m) above their base. Klyuchevskoy volcano in Kamchatka, Russia, is one of the highest in the world, which stands 15501.97 feet (4725 m) above sea level. Most composite volcanoes have a crater at the summit, which contains a central vent or a clustered group of vents. One essential feature about composite volcanoes is the conduit system, where the magma (molten rock material) from a reservoir deep in the Earth's crust rises to the surface. This type of volcano is built by the accumulation of materials erupted through the conduit, which increases in size as lava, cinders, and ash are added to its slopes due to multiple eruptions.

Composite volcanoes are the most dangerous types of volcanoes and are produced by intermediate composition magma or andesitic magma (silica ≈ 60%). Andes Mountains, Cascade Mountains (Mt. St Helens) and Ring of Fire in the Pacific Ocean are some examples. Composite volcanoes are smaller than shield volcanoes but larger than cinder cones and volcanic domes.

Shield Volcanoes

Shield volcanoes are built almost entirely of lava flow with low viscosity. The basaltic lava (silica ≈ 50%) gives these volcanoes their characteristics and style of eruption. The flow of low viscosity lava

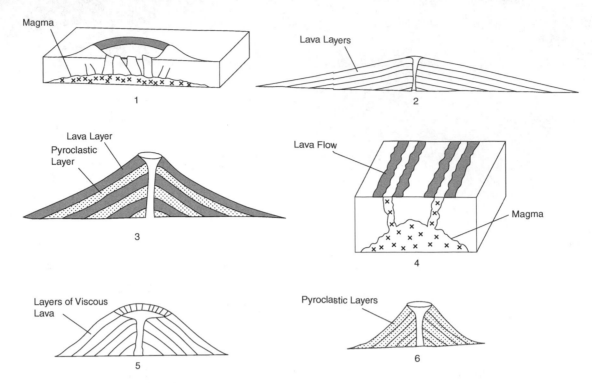

Figure 11.1 Different types of volcanoes and volcanic activity. The drawings are not to scale.

produces gentle slopes with low elevation. The base of the shield volcanoes are broad and rounded giving the shape of warrior's shield and this is why they are considered the largest volcanoes. Mauna Loa in Hawaii is one example.

Volcanic Domes

Smallest type of volcanoes that have steep slopes and are produced from viscous lava (rhyolitic lava where silica ≈ 70%), forming a thick, bulbous dome above and around a volcanic vent like the volcanic domes found inside the crater of Mt St Helens. Sometimes, these volcanoes start with explosive eruptions due to build up of gases that fade away with time.

Fissure Eruptions

Are referring to volcanic eruption from or within a fissure or series of fissures rather than through a central vent where basaltic lava flow (low viscosity) cover large areas as sheets or plates like the Columbia Plateau. Deccan Traps in India is another example.

Caldera

Caldera is a crater with diameter more than 1 km and they are produced by a combination of the explosion and collapse of the top of a volcanic cone. The vent looks like a very large, bowl-shaped volcanic depression whose horizontal dimension is much greater than its vertical dimension. Some calderas are larger than shield volcanoes. Yellowstone caldera complex is one of the largest calderas in the world, and it is made of three calderas; Henry's Fork, Island Park, and Yellowstone caldera.

_____3_____ Composite Volcano		_____4_____ Fissure Eruption
_____1_____ Caldera		_____6_____ Cinder Cone
_____5_____ Volcanic Dome		_____2_____ Shield Volcano

PART (B)

Referring to Figure (11.2), answer the following questions:

1.B)   Match the numbers in the figure with the specific feature or formation describing the volcanic features and igneous bodies.

1	Batholith	**8**	Volcanic Pipe
9	Crater	**5**	Sill
2	Dike	**3**	Lava Flow
4	Stock	**10**	Strato/Composite Volcano
6	Laccolith	**7**	Fissure Eruption (may develop in future)

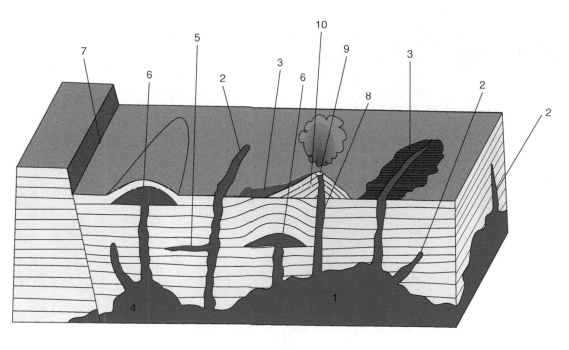

Figure 11.2

2.B)   List by name those features/formations above which are *intrusive*.

Batholith   Stock   dike   Volcanic   laccolith   sill
                                    pipe

3.B)   List by name those features/formations above which are *extrusive*.

Crater   lava flow   strato   fissure eruption

4.B) Which of the given features/formations probably consists of material of the largest grain size (phaneritic texture)? Why? *intrusive, the rate of cooling is small*

5.B) Which of the given features/formations probably consists of material of the smallest grain size (aphanitic texture)? Why? *extrusive the rate of cooling is slow*

PART (C)

Give the location of the following volcanoes/volcanic activity and put **X** in the proper space concerning volcanic type for the given volcanoes in Table (11.1).

Table 11.1

Name of Volcano	Location	Type of Volcano/Volcanic Activity					
		Shield	Composite	Cinder Cone	Volcanic Dome	Caldera	Fissure
Stromboli	Italy		X				
Mt. Pinatubo	Philipins		X				
Crater Lake	USA					X	
Mount Hood	USA		X				
Kilauea	Hawaii	X					
Paricutin	mexico			X			
Yellowstone	USA					X	
Mount Cotopaxi	equator		X				
Mount Saint Helens	USA		X				
Mount Rainier	USA		X				
Mauna Loa	pacific		X				
Mount Fuji	Japan		X				
Mount Shasta	USA		X				
Lakagigar or Laki	iceland						X
Mt. Etna	Italy		X				

PART (D)

You will be given the following:

1. Volcano plastic model #2
2. Volcano model legend. The legend is to explain the symbols for the volcano plastic model. Look at Figure (11.3).

Symbol	Meaning
	Conglomerate
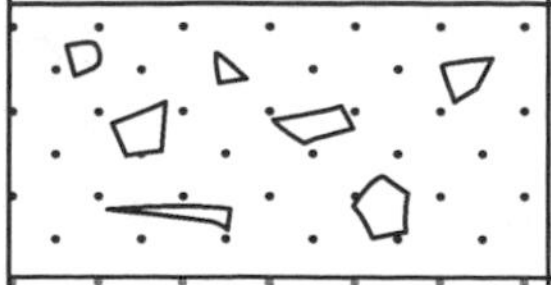	Breccia
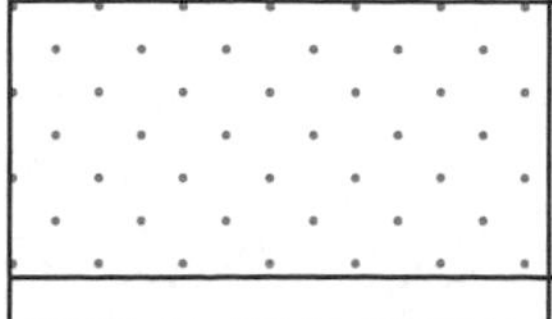	Sandstone (Shown as orange color)
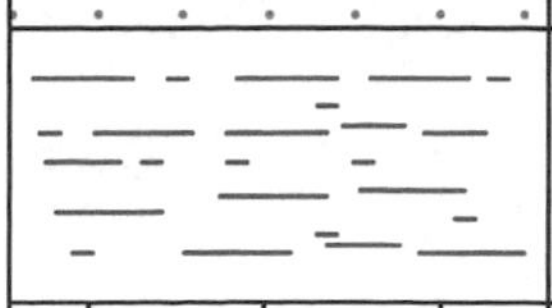	Shale (Shown as green/blue color)
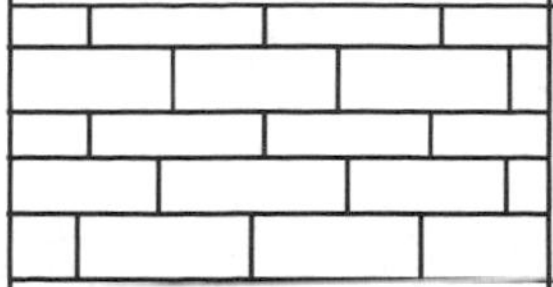	Limestone
	Magma
	Lava (Shown as brown color)
	Ash (Shown as gray color)

Figure 11.3

Fill Table (11.2) by trying to determine the name of the feature on the volcano plastic model using the terms given below based on their location and identification number; that is match the number of the feature using its coordinates with the terms given below:

Laccolith, Lava, Pyroclastic Flow, Crater, Lava Flow, Fault, Caldera, Dome, Lava Flow From Side Of Volcano, Old Volcano, Radial Drainage, Volcanic Ash, Small Cone In Caldera, Lava flow From Breached Crater, Radial Dike Outcrops, Conduit, Magma, Lake, Sill, Stratovolcano/Composite Volcano, Braided Stream, Series Of Small Cinder Cones, Breached Crater, Dike, Shield Volcano, Volcanic Neck.

Table 11.2

Number: Numbers are for features and are written in black and are shown on surface and sometimes shown on sides	Coordinates: Coordinates are for location of features and are written in blue and shown on sides, as letters and numbers and sometimes as letters only	Name of the feature:
21	G-10	
22	H-10	
23	H-10	
24	F-10	
25	F-14	
26	D-8	
27	E-11	
28	A-10	
29	C-11	
30	C-11	
31	D-11	
32	C-12	
33	C-14	
34	A-13	
35	A-13	
36	A-14	
37	H-14	
38	H-14	
39	A-10	
40	A-10	
41	A-10	
42	H-13	

PART (E)

With the aid of stereoscope, answer the following questions using the specific plates in the book titled "Aerial Stereograms: An introduction to geology, geography, conservation, forestry, and surveying using stereo photographs" by Harold R. Wanless and published by Hubbard Scientific.

Plate 50

1.E)   What type of volcano is plate (50) showing?

Cinder cone

2.E)   This type of volcano is made of what type of pyroclasts?

Cinders

3.E)   There is white narrow line going around the volcano from base to rim at crater. What does the white narrow line represent?

trail or road

4.E)   What are the grooves or gullies radiating down slope from crater represent?

drainage for rock fragments

5.E)   Are there any trees growing at slopes of this volcano? If yes where? Give coordinates.

yes

6.E)   Is the volcano active or not? If not, give two clues to support your answer.

no, because there are trees and a road

7.E)   What are the angular mounds near the base of volcano (to right)?

ancient lava flow

8.E)   Why is the lava flowing to the right of volcano?

steep slope

9.E)   Is this lava flowing from crater or base of volcano?

base of volcano

10.E)   If you are going to build a city close to volcano, where would be its location. Write the coordinates. *the left side away from the lava flow*

Plate 51

11.E)   What type of volcano is plate (51) showing? This question is referring to the dark small volcano at C.9-2.5.

12.E)   There are two lava flows, one is dark black and the other one is light gray. Which one is younger?

Plate 52

13.E)   Are there any roads seen in plate (52)? If answer is yes, give the coordinates.

14.E)   If there is/are road(s), is there something covering the road(s)? If yes, what is the nature of that thing?

15.E)   What is the white material found in the sea near the shore at lower part of stereogram?

16.E) From previous questions and answers, is this volcano showing signs of activity? If yes, what is/are this/these sign(s)?

Plate 54

17.E) What is the name of feature at D.0-2.8?

18.E) What does the feature at D.0-2.8 coordinates represent?

19.E) There is a bay at the top part of the feature located at D.0-2.8. How did this bay form?

20.E) At this specific location shown in plate (54), is erosion or deposition the dominant process?

21.E) What caused the three craters at A.9-1.9, B.1-2.1, and B.8-2.8 to be on same line?

22.E) Any vegetation detected in plate (54)?

Plate 56

23.E) How do caldera form?

24.E) Did Crater Lake form in caldera?

25.E) What is the feature located at B.6-1.9 called?

26.E) What does the light patch at A.8-2.6 represent?

27.E) Is the lake found at high or low elevation? Give one reason to support your answer.

28.E) Where did the water in the crater (lake) come from?

Plate 58

29.E) This crater is produced by meteorite impact. How can you differentiate between craters produced by meteorite impact and craters produced by volcanic activity?

30.E) Is the rim (wall) of crater at top of stereogram higher or lower than the rim (wall) at the bottom of stereogram?

31.E) Based on your answer to question (30.E), does that give you a clue to which direction the meteorite struck the ground? Was the direction from bottom to top or top to bottom of the stereogram?

32.E) Did the meteor hit the ground, with 90° angle (directly above the ground) or with lower angle?

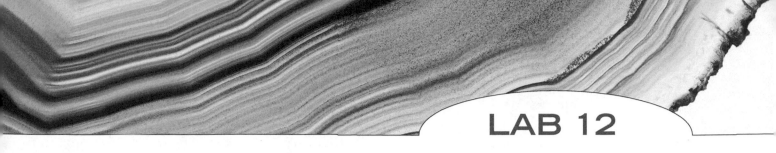

HISTORICAL GEOLOGY

PART (A): RELATIVE DATING

1.A)   Look at the different images of Figure (12.1). Starting from image (1), write down the geological history (events) in order starting from the oldest (bottom) to the youngest (top) in blanks (spaces) next to each image. Use legend supplied below.

Symbol	Meaning
	Conglomerate
	Sandstone
	Siltstone
	Shale
	Limestone
	Contact Metaphormism/Baking Area
	Magma/Igneous Rock/Igneous Body

Legend to be used for images of Figure (12.1) showing the different symbols used and their meaning.

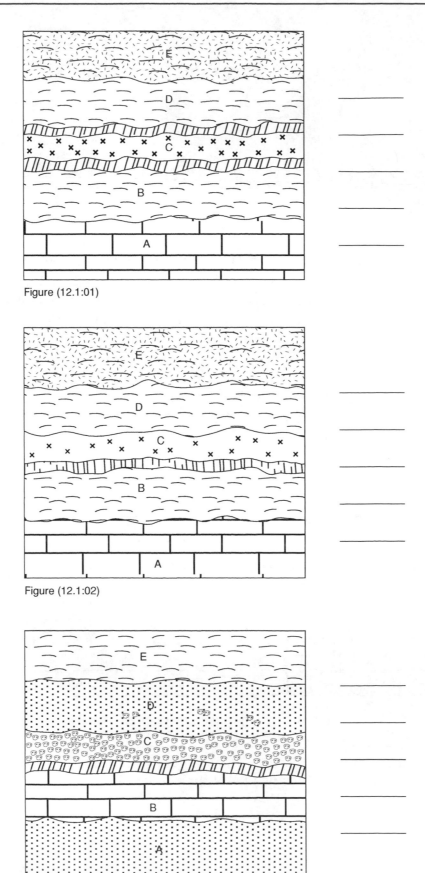

Figure (12.1:01)

Figure (12.1:02)

Figure (12.1:03)

Figure 12.1 (*Continued*)

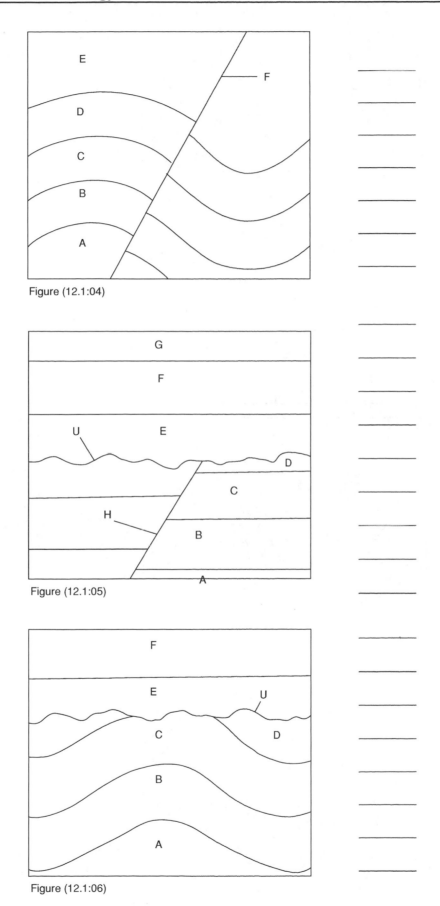

Figure (12.1:04)

Figure (12.1:05)

Figure (12.1:06)

Figure 12.1 (*Continued*)

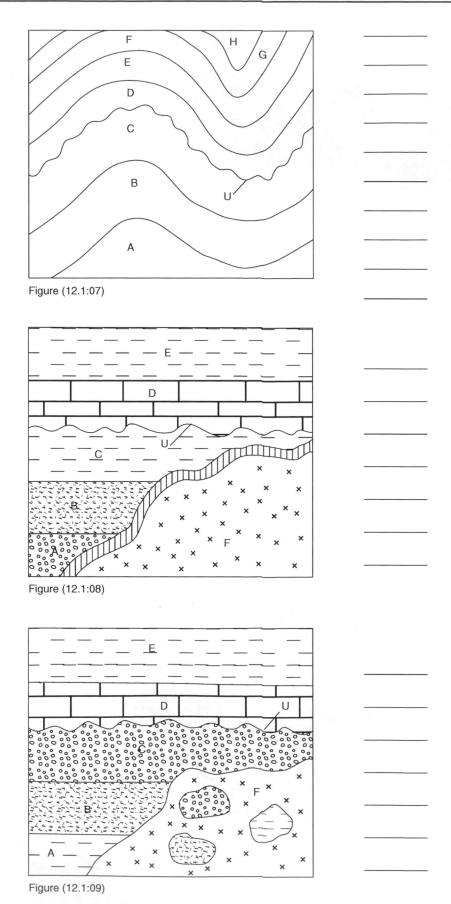

Figure (12.1:07)

Figure (12.1:08)

Figure (12.1:09)

Figure 12.1 (*Continued*)

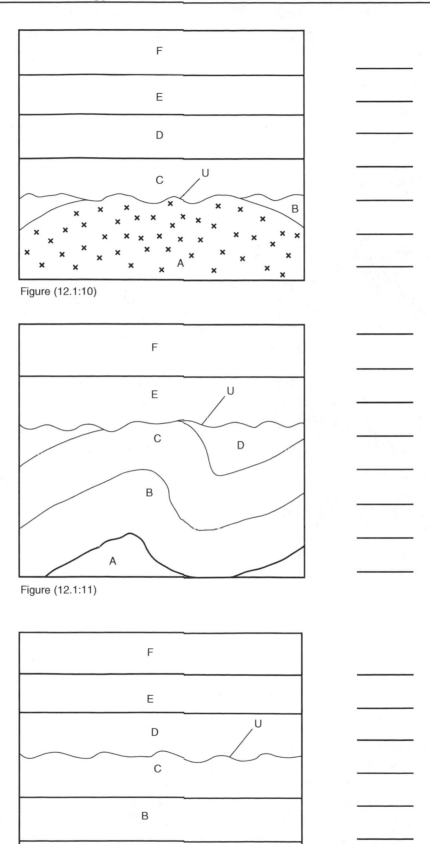

Figure (12.1:10)

Figure (12.1:11)

Figure (12.1:12)

Figure 12.1

2.A)   Put the events of Figure (12.2) in order starting from the oldest to youngest, in the following spaces.

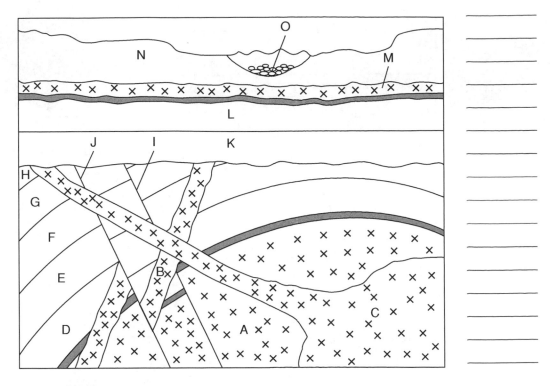

Figure 12.2

PART (B): CORRELATION

Correlation is relating rocks or group of rocks that formed at two different locations (or more). By using correlation, it is possible to find out if rocks at different locations formed at the same time.

The correlation could be based on lithology and time. Correlation could have more than one possible solution. Always make list of solutions and after that select the simplest solution and simple here means the solution that explains the most situations with the least assumptions.

Generally speaking, when correlating outcrops, it is a good practice to start with unconformities, key beds and chronostratigraphic units.

Lithology

The correlation used here is based on type of rocks and their characteristics like mineral composition, texture, color, and structure so time is not involved. Lithostratigraphic unit is layer distinguished from upper and lower layer based on lithology and not time and these distinguished lithological layer of rock(s) are known as lithostratigraphic units and are divided into groups, formations, members, and beds. For short distances, the lithostratigraphic correlation could be same as chronostratigraphic correlation, however for medium distances they are not the same and for long distances only chronostratigraphic correlation could be used.

For correlation that is based on lithology it is better to use "key beds" like coal seam, volcanic ash, and meteorite deposits.

1.B) Try to correlate locations (A) and (B) for Figure (12.3).

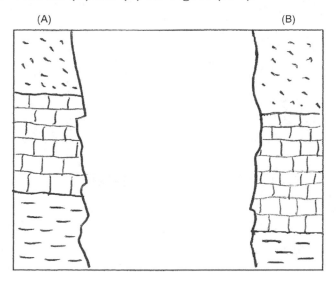

Figure 12.3

2.B) What type of correlation you used for figure (12.3), that is your correlation was based on what?

3.B) Try to correlate the outcrops of Figures (12.4)–(12.7) based on lithology. Pay attention to "pinch out" (layer becomes thin and then ends with distance) and "change of facies" (change of depositional environment).

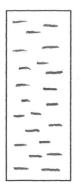   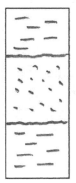

Figure 12.4

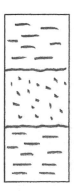   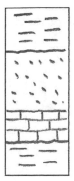

Figure 12.5

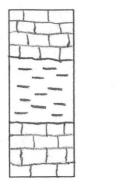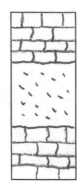

Figure 12.6

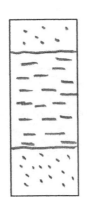

Figure 12.7

4.B) Correlation of the same outcrops could be done more than one way because the interpretation used for correlation could be based on more than one logic. Try to correlate Figure (12.8) based on lithology and notice that the three parts (1, 2, 3) of Figure (12.8) are all identical (same).

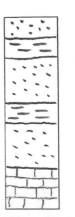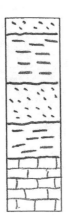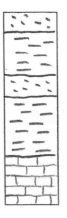

Figure 12.8: 1

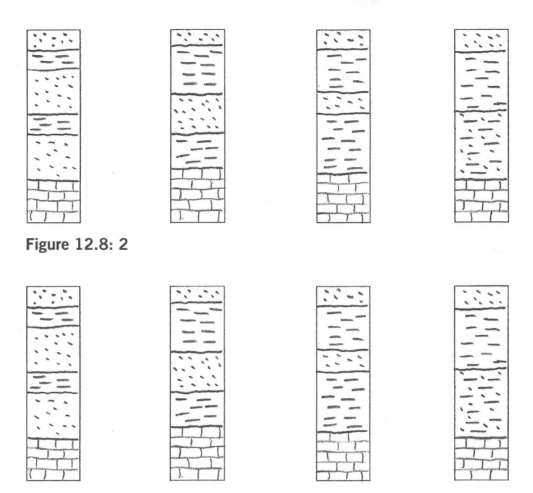

Figure 12.8: 2

Figure 12.8: 3

5.B) Could correlation between two locations be done based on lithology only? Explain your answer.

6.B) Now look at Figure (12.9). This figure may give the answer for question (5.B). Can you correlate the two locations at (A) and (B)?

7.B) If the answer to question (6.B) is no, explain why?

8.B) If the answer to question (6.B) is yes, explain why?

9.B) Assume that correlation is possible for the two locations of Figure (12.9), this correlation will be based on what?

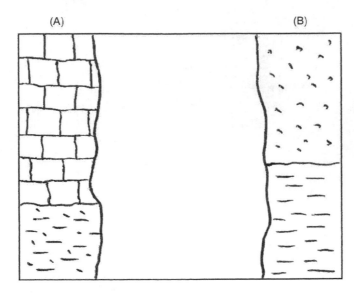

Figure 12.9

Time (chronology)

This correlation is based on radioactive isotopes or fossils and the correlation is used for long distances. There are two time stratigraphic units:

a. Chronostratigraphic Units or time stratigraphic units: refers to rock(s) that formed at the same specific time so type of rock is not important and are divided into erathem, system, series, stage, and chronozone. So Permian system is referring to all rocks that formed during the Permian time. The time span is known usually using fossils (biostratigraphy) and sometimes by using radioactive isotopes. It is interesting to know that most of the fossils and most of the sedimentary rocks can't be dated using radioactive isotopes. Furthermore, if fossils are used it should be specific type of fossils known as index fossils and in this case biostratigraphic units are rocks or layers recognized and distinguished based on fossils only. The basic unit for biostratigraphic units is biozone.

b. Geochronological units: refers to divisions of time and they are divided to eon, era, period, epoch, age. For example, Permian period is referring to time span from 250 to 300 million years.

We will use fossils only for this type of correlation

It is possible that rocks at two locations that formed at the same time will have different lithology and the cause of this difference in lithology could be related to change of environment of formation or deposition. In this case, correlation based on lithology is not enough and use of fossils is necessary. Caution should be used when using fossils as it will be seen and if fossils are used for correlation then index fossils only must be used.

10.B) Figure (12.10) shows four outcrops for layers numbered from 1 to 4. In some layers of the outcrops there are index fossils shown as symbols like star, double circle, triangle, and plus sign.

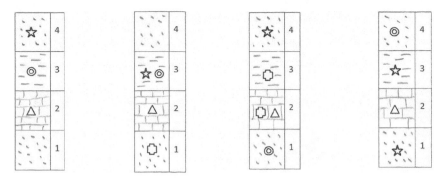

Figure 12.10

There are four index fossils, which index fossil out of the four should be used and why?

11.B)   Try to correlate the two outcrops of Figure (12.11) based on fossils. Notice that the symbols in layers like double circle, plus sign, triangle, and star represent index fossils.

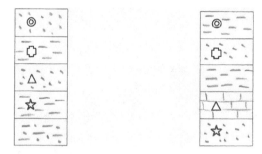

Figure 12.11

12.B)   Can Figure (12.12) be correlated on lithology?

13.B)   Can Figure (12.12) be correlated based on fossils?

14.B)   It is possible to correlate the two locations (outcrops) at (A) and (B) of Figure (12.12). Say why.

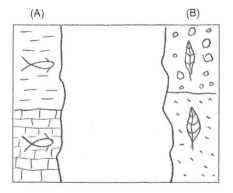

Figure 12.12

15.B)   Try to correlate locations (A) and (B) for Figure (12.13).

16.B)   If correlation for Figure (12.13) was possible, in your opinion this correlation was based on what?

17.B)   If correlation for Figure (12.13) was not possible, in your opinion why this correlation was not possible?

18.B)   In your opinion how can we explain Figure (12.13)? Do you see difficulty to correlate the two locations and are index fossils and lithology enough to make correlation?

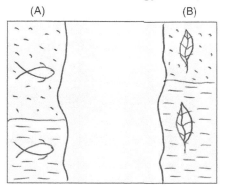

Figure 12.13

From the previous figures, we see that index fossils do help to correlate between different locations and when using index fossils we need to be careful and we need other tools to help in the process of correlation and this is why fossil range and fossil assemblage are introduced.

Fossil range is referring to time interval between first and last appearance of a fossil.

Fossil assemblage is overlap of fossils because more than one type of organism (fossil) lived at the same time.

19.B)   Using Figure (12.14), try to draw fossil range and fossil assemblage using arrows of different colors, if possible. Note that layers are numbered and the index fossils are shown as symbols like star, double circle, and triangle.

Layer Number	Fossil Type	Fossil Range	Fossil Assemblage
6	☆		
5	☆		
4	☆ ◎		
3	◎ △		
2	△		
1	△		

Figure 12.14

PART (C): ABSOLUTE DATING

Assume that a rock contains radioactive isotope with a half life equal to 2000 years. Furthermore, assume that the ratio of the parent atoms to daughter atoms is 1:3. How is it possible to find the age of the rock?

It is known that half life is the time taken for half of the atoms to decay and from this definition we know that after 2000 years, half of the radioactive isotope (the original isotope called parent isotope) will decay to another isotope (called the daughter isotope). This will continue until all the parent radioactive isotope changes completely to a daughter isotope. Look at Table (12.1).

Table 12.1

Number of Parent Atoms	Number of Daughter Atoms	Ratio Form	Fraction Form
50	50	1:1	1/2 + 1/2
25	75	1:3	1/4 + 3/4
12.5	87.5	1:7	1/8 + 7/8
6.25	93.75	1:15	1/16 + 15/16
3.125	96.875	1:31	1/32 + 31/32

From Table (12.1) note that ratio form was obtained by writing one on the left side and then dividing daughter atoms by parent atoms to get the right side of the ratio so 50/50 = 1, 75/25 = 3, 87.5/12.5 is 7 and so on.

The fraction ratio was obtained using this method:

For the left side always put 1 on the numerator. The dominator was obtained by simply adding the ratio rate. For example, $1 + 1 = 2$, $1 + 3 = 4$, and so on.

The right side of fraction form was obtained by putting the addition of the ratio form as the dominator and figuring out the numerator that will be added to the right side to get one. For example $1/2 + X/2 = 1$, so X must be 1. Same thing $1/4 + X/4 = 1$, so X must be 3 and so on.

Now after knowing how to express the ratio form and fraction form for parent and daughter atoms, we can determine the age of the rock from the formula:

$$\text{Age} = \text{Number of half lives} \times \text{half life}$$

Half life is given and the problem is how to obtain number of half lives. Number of half lives can be obtained from Table (12.1) and the decay curve. Look at Figure (12.15).

Since the ratio of parent atoms to daughter atoms is 1:3 then using the above table we see that only 1/4 of the parent atoms stayed as it is and did not decay and 1/4 = 0.25. Now plot 0.25 (which is 25%) on the Y-axis of the decay curve and make a horizontal line that will intersect the curve and from there (the intersection) drop a vertical line to the X-axis to give you the number of half lives. In this case, the number of the half lives = 2, so the age of the rock = number of half lives × half life

$$= (2) \times (2000) = 4000 \text{ years}$$

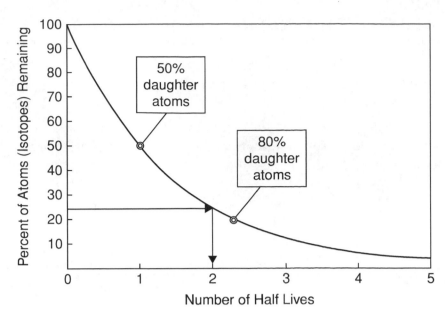

Figure 12.15 Decay Curve

Use Table (12.1) and Figure (12.15) to answer the following questions.

1.C)   Assume that you have 100 atoms of radioactive isotope and the half life for this radioactive isotope is 5000 years. After four half lives, there will be no atoms left of the original (parent) isotope.

Say if the previous statement is true or false and try to support your answer with logical reasoning.

2.C)   If specific amount of parent atoms (unknown) gave only 15/16 of the daughter atoms then how many half lives passed?

3.C)   Assume that the ratio of parent atoms to daughter atoms is 1:15 and the radioactive isotope inside the fossil/rock has a half life of 10,000 years. Determine the age of the sample. Show your calculations.

4.C)   What would be the age of the fossil/rock if it had parent to daughter ratio of 1:13 (use same half life of 10,000 years). Show your calculations.

5.C)   If 75% of the parent isotope decayed, how many half lives passed?

PART (D): FOSSIL HUNTING

Fossil hunting is not only interesting hobby but also educational and rewarding.

In this activity, you will use a guide book and fossils box supplied by your instructor and the guide book is used to identify fossils and their age. The following steps are used to fill Table (12.2).

For those who want to use the guide book for fossil hunting in general and not only with this lab book, the name of the guide book is given:

National Audubon Society: Field Guide to Fossils (North America).

A. First, you need to determine the age of the rocks that the fossil(s) (you have or own) belong to and that depends on the state you are living in. So first find ages of rocks in that specific state. To do that:

1.   Turn to index map of North America on page 104 (colored states map).
2.   Illinois State is shown as light green (not dark or medium green).
3.   Open thumb tabs pages that show light green. Thumb tabs are the next pages after the index map of North America.
4.   After finding the light green thumb tab page, you will find map showing three states. Illinois is one of them. Now determine the ages of rocks in Illinois State by looking at the symbols that represent the ages of rocks. Look for these symbols carefully (for Illinois State). You may need magnifying glasses. You could find the following (or more):

P_2, S, D, €

Now from p. 102, the above symbols mean the following ages (periods):

P_2 means Permian
S means Silurian
D means Devonian
€ means Cambrian

B. Now bring the fossil that you found (or have) and compare it to the drawings that start after p. 149 (these pages are black and the drawings are white). Try to find the closest drawing that looks like your fossil. These pages start after p. 149. Page 149 is the Thumb Tab Guide.

1. Assume that your fossil looks like the "Scallop-shaped Fossils," you will find next to the "Scallop-shaped Fossils" there is number and this number is Plate Number 34-102.

2. You go to plates 34 to 102 (these are colored plates of real fossils and not drawings) and select the one that is closest to your sample. Let us assume that plate 35 (Chlamys) is the closest plate that looks like your fossil sample, then you see if that is the right one or not by going to p. 465 (this number is written next to Chlamys) and on p. 465 you find all the information about that fossil that resembles your sample, like age and phylum. If the information is right (your fossil sample age is same as the age of the colored fossil plate) then you got it and if not then you have to look for the right one again by starting from step (B) above. Practice since practice makes perfect.

Table 12.2

Fossil Number	Phylum	Age Range	Fossil Name

A: GEOLOGIC TIME SCALE

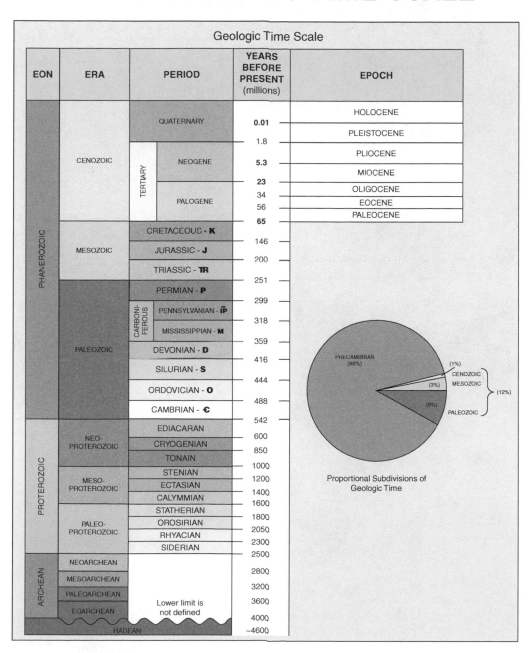

EON	ERA	PERIOD			YEARS BEFORE PRESENT (millions)	EPOCH
PHANEROZOIC	CENOZOIC		QUATERNARY		0.01	HOLOCENE
						PLEISTOCENE
					1.8	
		TERTIARY	NEOGENE			PLIOCENE
					5.3	MIOCENE
					23	OLIGOCENE
			PALOGENE		34	EOCENE
					56	
					65	PALEOCENE
	MESOZOIC	CRETACEOUS - **K**			146	
		JURASSIC - **J**			200	
		TRIASSIC - **TR**			251	
	PALEOZOIC	PERMIAN - **P**			299	
		CARBONIFEROUS	PENNSYLVANIAN - **IP**		318	
			MISSISSIPPIAN - **M**		359	
		DEVONIAN - **D**			416	
		SILURIAN - **S**			444	
		ORDOVICIAN - **O**			488	
		CAMBRIAN - **C**			542	
PROTEROZOIC	NEO-PROTEROZOIC	EDIACARAN			600	
		CRYOGENIAN			850	
		TONAIN			1000	
	MESO-PROTEROZOIC	STENIAN			1200	
		ECTASIAN			1400	
		CALYMMIAN			1600	
	PALEO-PROTEROZOIC	STATHERIAN			1800	
		OROSIRIAN			2050	
		RHYACIAN			2300	
		SIDERIAN			2500	
ARCHEAN	NEOARCHEAN				2800	
	MESOARCHEAN				3200	
	PALEOARCHEAN				3600	
	EOARCHEAN	Lower limit is not defined			4000	
HADEAN					~4600	

PRECAMBRIAN (88%)
(1%) CENOZOIC
(3%) MESOZOIC
(8%) PALEOZOIC
(12%)

Proportional Subdivisions of Geologic Time

B: PHYSIOGRAPHY MAP OF USA

© Intrepix/Shutterstock.com

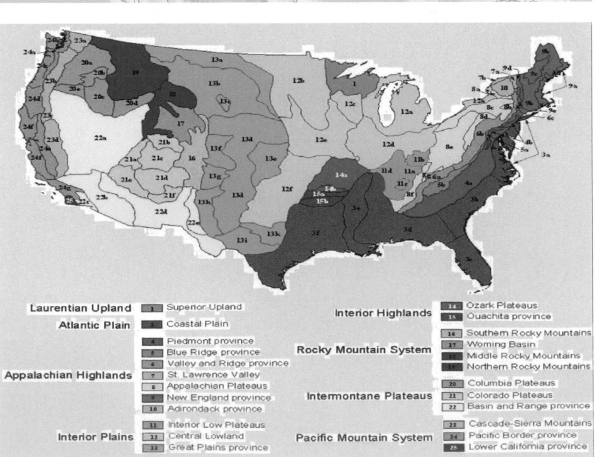

Laurentian Upland	**1**	Superior Upland
Atlantic Plain	**3**	Coastal Plain
	4	Piedmont province
	5	Blue Ridge province
	6	Valley and Ridge province
Appalachian Highlands	**7**	St. Lawrence Valley
	8	Appalachian Plateaus
	9	New England province
	10	Adirondack province
	11	Interior Low Plateaus
Interior Plains	**12**	Central Lowland
	13	Great Plains province

Interior Highlands	**14**	Ozark Plateaus
	15	Ouachita province
	16	Southern Rocky Mountains
Rocky Mountain System	**17**	Woming Basin
	18	Middle Rocky Mountains
	19	Northern Rocky Mountains
	20	Columbia Plateaus
Intermontane Plateaus	**21**	Colorado Plateaus
	22	Basin and Range province
	23	Cascade-Sierra Mountains
Pacific Mountain System	**24**	Pacific Border province
	25	Lower California province

Physiographic Map Reference: Fenneman, Nevin M., 1946, *Physical Divisions of the United States:* U.S. Geological Survey, scale 1:7,000,000.

C: GEOLOGIC MAP OF NORTH AMERICA

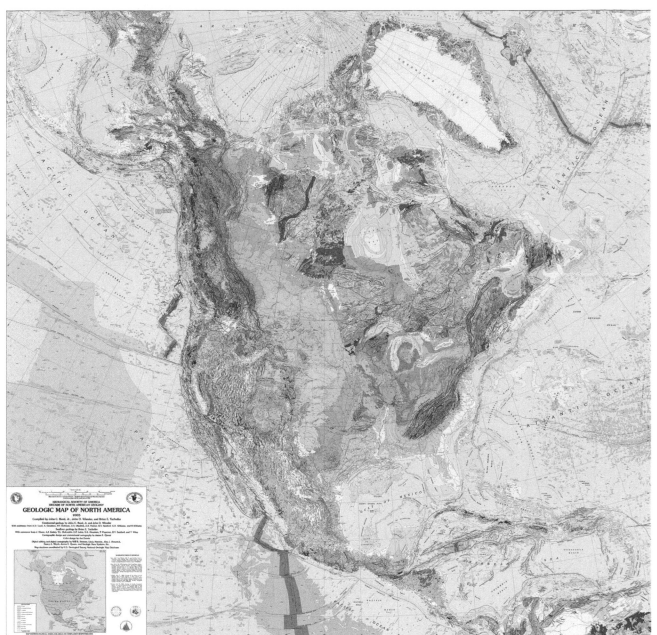

D: METRIC AND ENGLISH UNITS

Selected Prefixes Used in the Metric System

Prefix	Abbreviation	Meaning	Example	
Giga	G	10^9	1 gigameter (Gm)	$= 1 \times 10^9$ m
Mega	M	10^6	1 megameter (Mm)	$= 1 \times 10^6$ m
Kilo	k	10^3	1 kilometer (km)	$= 1 \times 10^3$ m
Deci	d	10^{-1}	1 decimeter (dm)	$= 0.1$ m
Centi	c	10^{-2}	1 centimeter (cm)	$= 0.01$ m
Milli	m	10^{-3}	1 millimeter (mm)	$= 0.001$ m
Micro	μ[a]	10^{-6}	1 micrometer (μm)	$= 1 \times 10^{-6}$m
Nano	n	10^{-9}	1 nanometer (nm)	$= 1 \times 10^{-9}$ m
Pico	p	10^{-12}	1 picometer (pm)	$= 1 \times 10^{-12}$ m
Femto	f	10^{-15}	1 femtometer (fm)	$= 1 \times 10^{-15}$ m

[a]This is the Greek letter mu (pronounced "mew").

For example:

1 kilometer (km) = 0.62 mile (mi)

1 kilometer (km) = 3280.8 feet (ft)

1 meter (m) = 3.28 feet (ft)

1 centimeter (cm) = 0.39 inch (in)

1 millimeter (mm) = 0.039 inch (in)

1 inch (in) = 2.54 centimeters (cm)

1 inch (in) = 25.4 millimeters (mm)

1 foot (ft) = 0.30 meter (m)

1 yard (yd) = 0.91 meter (m)

1 yard (yd) = 0.00091 kilometer (km)

1 mile (mi) = 1.61 kilometers (km)

The Fraction would be:

$$\frac{1 \text{ kilometer (km)}}{0.62 \text{ mile (mi)}} \quad \frac{\text{(numerator)}}{\text{(denominator)}}$$

Length Conversion

kilometer (km) → meter (m)

meter (m) → centimeter (cm)

centimeter (cm) → millimeter (mm)

Mass Conversion

kilogram (kg) → gram (g)

gram (g) → milligram (mg)

Volume Conversion

kiloliter (kl) → liter (l)

liter (l) → milliliter (ml)